TIME-FREQUENCY ANALYSIS AND SYNTHESIS OF LINEAR SIGNAL SPACES

Time-Frequency Filters, Signal Detection and Estimation, and Range-Doppler Estimation

TIME-FREQUENCY ANALYSIS AND SYNTHESIS OF LINEAR SIGNAL SPACES

Time-Frequency Filters, Signal Detection and Estimation, and Range-Doppler Estimation

by

Franz Hlawatsch
Vienna University of Technology

KLUWER ACADEMIC PUBLISHERS
Boston / Dordrecht / London

Distributors for North America:
Kluwer Academic Publishers
101 Philip Drive
Assinippi Park
Norwell, Massachusetts 02061 USA

Distributors for all other countries:
Kluwer Academic Publishers Group
Distribution Centre
Post Office Box 322
3300 AH Dordrecht, THE NETHERLANDS

Library of Congress Cataloging-in-Publication Data

A C.I.P. Catalogue record for this book is available
from the Library of Congress.

Contents

Preface

It is now nearly 20 years ago that the Wigner distribution (also known as Wigner-Ville distribution) was given a prominent place in the signal processing community's spectrum of research interests. Recognition was given almost immediately for the high potential of the Wigner distribution as a tool for displaying and analyzing signal characteristics in the time-frequency plane that are observable only indirectly from the temporal or spectral description of the signal. This assessment was soundly supported by the long list of desirable mathematical properties (such as correct marginal and support properties) satisfied by the Wigner distribution. Moreover, unlike, for instance, the spectrogram, the resolution of the Wigner distribution is not limited by the temporal or spectral properties of a window, simply because such a window is not required in the definition of the Wigner distribution.

On the other hand, due to its high resolution and its quadratic nature, the Wigner distribution locally exhibits oscillations between positive and negative values to an extent that smoothing, in a certain sense in accordance with the Heisenberg uncertainty principle, is necessary to make it non-negative everywhere. As a consequence, a pointwise interpretation of the Wigner distribution as a true probability density function is not possible. This circumstance has kept the discussions in the signal processing community quite animated, yielding all sorts of modified Wigner distributions through the design of smoothing kernels and distributions involving the signals in a non-quadratic way. Thus, to a large extent, the interest and research effort of the community has been directed at redefining and re-interpreting the Wigner distribution, with the purpose of obtaining a clear and useful display and analysis tool.

Besides being a useful display and analysis tool, the Wigner distribution can be of great help in the design of time-frequency filtering methods. In this book such methods are designed for the enhancement, decomposition, estimation and detection of noisy deterministic and stochastic signals. These methods are based on the assumption that there exists information about the regions in the time-frequency plane where the signals and noise manifest themselves. The central idea here is that the linear operators involved in the solutions of many signal processing problems can be compressed to finite-dimensional linear spaces, the dimensions of which are approximately equal to the area of the region in the time-frequency plane to which the signals are supposed to be restricted. As a consequence, a major obstruction, viz. the composition and inversion of these operators, is removed since the linear spaces on which the operators have to be considered are finite dimensional. Besides that, formulation of the available information in terms of regions in the time-frequency plane has the advantage of yielding a direct interpretation of the effect of adding or deleting information and of changing parameters in the signal processing methods. In a sense, the information and the signal processing tasks are brought to life in the time-frequency plane.

In this approach to designing signal processing methods, the concept of the Wigner distribution of a linear signal space is a crucial notion. This concept can be defined simply as the sum of all Wigner distributions of the elements of an arbitrary orthonormal base of the linear space. The key step then consists of finding for a given region in the time-frequency plane that linear space whose Wigner distribution yields the best approximation to the indicator function of the considered region. It turns out that this amounts to rounding to 0 or 1 the eigenvalues of the linear operator that is associated with the indicator function via Weyl's correspondence principle.

The idea of defining the Wigner distribution of a linear space can be extended to more general quadratic signal representations than the Wigner distribution. This is worked out in this book for the ambiguity function, which is well-known in radar analysis. The sort of applications of the concept obtained in this way is quite different from the ones earlier mentioned, since Wigner distributions aim at describing energy distributions over time and frequency, whereas ambiguity functions are meant for revealing correlative structures. The concept of ambiguity function of a linear space thus appears to be quite instrumental in solving the problem of optimal design of sets of narrowband pulses to be used in the estimation of the range and Doppler shift of a slowly fluctuating point target.

The book consists of 9 chapters. A brief description of each of them can be found in Section 1.3. Chapters 2-6 form a rather coherent part of the book, with Chapter 4 being the cornerstone of the approach of designing signal processing methods for which time-frequency considerations provide the guiding principles. In this Chapter 4 the problem of synthesizing linear signal spaces is solved: with a given region in the time-frequency plane there is associated the linear signal space whose Wigner distribution is closest to the indicator function of the region. The results of Chapter 4 find ample application in Chapters 5 and 6, in which linear signal spaces with prescribed time-frequency localization are required for a variety of signal processing problems. Chapters 7-8 form a second coherent part of the book. The notion of ambiguity function of a linear signal space is introduced and applied to the problem of optimally designing a set of radar pulses. Here, "optimal" refers to minimum Cramér-Rao lower bound and maximum global accuracy of the maximum likelihood multipulse estimator for range and Doppler shift.

The prerequisites for reading the book are kept to a minimum. Aside from having a general interest in and insight into time-frequency analysis, some basic knowledge of linear algebra and Fourier analysis generally suffices. This is perhaps not so for some parts of Chapters 6 and 8, in which somewhat more advanced elements from statistical estimation and detection theory are used.

In order to advance an accessible presentation, all concepts are amply justified and illustrated by examples and pictures. Also, the theorems are, as a rule, directly followed by interpretations and applications, while their proofs are relegated to appendices when they do not contribute to a deeper understanding of the matter.

Finally a word about mathematical rigour. The results and notions of a more algebraic nature, as well as those concerning statistical estimation/detection theory and the time-frequency localization of linear signal spaces, lend themselves to a mathematically rigourous treatment without restraining the clarity of the exposition. This is not so for notions like essential support, oscillatory, underspread, simple and sophisticated spaces; these are treated in a more intuitive way. The reason for this is that, although some progress has been made in recent years, mathematical results on the supports and value distributions of Wigner distributions are rather scarce, and often so cumbersome to use that they would disturb the delicate balance between the element of engineering intuition and that of mathematical precision pursued in the exposition.

A. J. E. M. JANSSEN

Philips Research Laboratories Eindhoven

Acknowledgments

It is a pleasure to gratefully acknowledge the substantial contributions that several people have made to this work. I would like to specifically thank

- O. Univ.-Prof. Dr. W. Mecklenbräuker for having set the stage and for guidance all along the way;

- Dipl.-Ing. W. Kozek for many ideas that have substantially influenced and enriched this work;

- Dipl.-Ing. H. Kirchauer for his detailed analysis of the signal estimation schemes discussed in Section 6.1;

- Dipl.-Ing. G. Matz for working out much of the material on signal detection in Section 6.2, and for help regarding the layout;

- Dr. G. S. Edelson and Dipl.-Ing. P. Podlucki for important contributions to the analysis of the range-Doppler estimator proposed in Chapter 8;

- Dr. A. J. E. M. Janssen for writing the preface and for providing valuable insight concerning Laguerre functions as well as many other useful hints;

- Dipl.-Ing. U. Trautwein, Ing. B. Wistawel, Dipl.-Ing. K. Vavrina, Dipl.-Ing. P. Podlucki, Dipl.-Ing. M. Linsbauer, Dipl.-Ing. W. Kozek, and Dipl.-Ing. H. Fitz for contributing simulations, plots, and drawings;

- Mr. R. W. Holland, Jr., of Kluwer for his interest in this book project and for the excellent cooperation during its realization.

Finally, I am deeply grateful to my wife Renate for her understanding, patience, and support.

TIME-FREQUENCY ANALYSIS AND SYNTHESIS OF LINEAR SIGNAL SPACES

Time-Frequency Filters, Signal Detection and Estimation, and Range-Doppler Estimation

1   INTRODUCTION AND OUTLINE

Linear signal spaces are of great importance in signal and system theory, communication theory, and modern signal processing. Traditionally, a linear signal space is considered to be a more or less abstract mathematical concept. In this work, however, a time-frequency (TF) analysis of linear signal spaces is proposed whereby linear signal spaces are represented as surfaces extending over a joint TF plane. This new viewpoint is based mathematically on specific *TF representations of linear signal spaces* that extend well-known TF representations of signals. Thus, this work provides an extension of the TF analysis of signals to linear signal spaces.

The first goal of such a TF analysis is a *visualization* of linear signal spaces. Moreover, we shall show how this approach can also be used to *synthesize* linear signal spaces with specified TF localization properties. This, in turn, provides a powerful new method for the design of *TF filters* that pass or suppress signal components located in specified TF regions. Various applications in statistical signal processing (signal estimation, signal detection, and parameter estimation) will be discussed as well.

This introductory chapter is organized as follows. Section 1.1 provides a brief review of linear signal spaces. Section 1.2 reviews two fundamental TF signal representations on which this work is based, namely, the Wigner distribution and the ambiguity function. Finally, an outline of the remaining chapters and a summary of major results are provided in Section 1.3.

1.1 Linear Signal Spaces

Both for signal and system theory in general and for the formulation of modern signal processing algorithms, linear signal spaces and the associated concepts of orthogonal projections and orthonormal bases are of fundamental relevance. A few examples may suffice to illustrate the importance of these concepts:

- An idealized lowpass or bandpass filter can be interpreted as an orthogonal projection operator on a linear space of suitably band-limited signals [Naylor and Sell, 1982].

- The operations of sampling a band-limited signal and reconstructing the signal from its samples can be interpreted as an expansion of the signal into an orthonormal basis (the basis of shifted sinc functions) spanning the associated linear space of band-limited signals [Naylor and Sell, 1982].

- In the field of digital communications, the concept of linear signal spaces is fundamental to both the derivation of optimal receivers and a deeper understanding of the performance of these receivers for various types of modulation [Wozencraft and Jacobs, 1965, Lee and Messerschmitt, 1994, Lafrance, 1990].

- Linear signal spaces and orthogonal projections are key concepts for least-squares optimization [Lawson and Hanson, 1974], adaptive filters [Honig and Messerschmitt, 1984, Haykin, 1991], and statistical signal processing problems like detection and estimation [Franks, 1969, Scharf, 1991, Picinbono, 1980].

More generally, the concept of linear signal spaces is indispensable for a deeper understanding of *linear operators* [Franks, 1969, Naylor and Sell, 1982, Luenberger, 1969] or, equivalently, linear transformations, systems, or filters. Linear operators are ubiquitous in the theory and practice of signal processing.

A *linear signal space* $\mathcal{X}$ is a collection of signals $x(t)$ such that any linear combination $c_1 x_1(t) + c_2 x_2(t)$ of two elements $x_1(t) \in \mathcal{X}$ and $x_2(t) \in \mathcal{X}$ is again an element of $\mathcal{X}$ [Franks, 1969, Naylor and Sell, 1982]. In this work, we shall consider subspaces of the space $\mathcal{L}_2(\mathbb{R})$ of finite-energy signals, so that $\mathcal{X}$ is equipped with inner product[1] $\langle x_1, x_2 \rangle = \int_t x_1(t)\, x_2^*(t)\, dt$ and norm $\|x\| = \langle x, x \rangle^{1/2} = \left[\int_t |x(t)|^2\, dt \right]^{1/2}$.

---

[1] All integrals and sums go from $-\infty$ to ∞ unless specified otherwise.

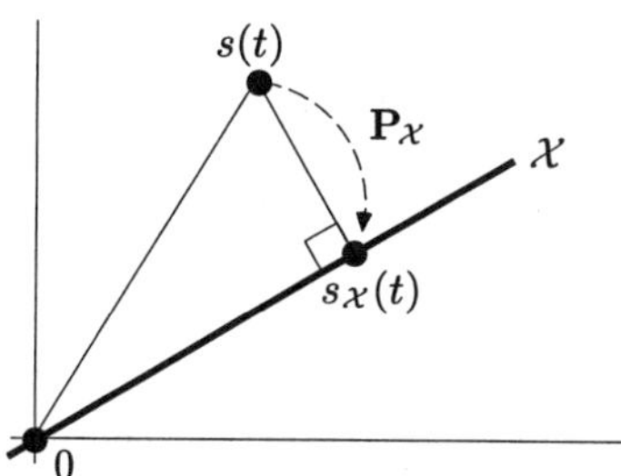

Figure 1.1.   Orthogonal projection of a signal $s(t)$ on a space $\mathcal{X}$. The space is symbolized by a straight line passing through the origin.

The *orthogonal projection* $s_{\mathcal{X}}(t) \in \mathcal{X}$ of a signal $s(t) \in \mathcal{L}_2(\mathbb{R})$ onto $\mathcal{X}$ can be written as

$$s_{\mathcal{X}}(t) = (\mathbf{P}_{\mathcal{X}}s)(t) = \int_{t'} P_{\mathcal{X}}(t,t')\, s(t')\, dt' \tag{1.1}$$

where $\mathbf{P}_{\mathcal{X}}$ denotes the orthogonal projection operator onto $\mathcal{X}$ and $P_{\mathcal{X}}(t,t')$ is its kernel. Equivalently,

$$s_{\mathcal{X}}(t) = \sum_{k=1}^{N_{\mathcal{X}}} s_k\, x_k(t) \qquad \text{with} \qquad s_k = \langle s, x_k \rangle , \tag{1.2}$$

where $\{x_k(t)\}_{k=1}^{N_{\mathcal{X}}}$ is an orthonormal basis of $\mathcal{X}$ and $N_{\mathcal{X}}$ is the dimension of $\mathcal{X}$. The orthogonal projection operator can be interpreted as a linear, time-varying system (filter). It is idempotent $(\mathbf{P}_{\mathcal{X}}^2 = \mathbf{P}_{\mathcal{X}})$ and self-adjoint $(\mathbf{P}_{\mathcal{X}}^{+} = \mathbf{P}_{\mathcal{X}}$, where $\mathbf{P}_{\mathcal{X}}^{+}$ denotes the adjoint of $\mathbf{P}_{\mathcal{X}}$ [Franks, 1969, Naylor and Sell, 1982]), and it can be expressed in terms of any orthonormal basis $\{x_k(t)\}_{k=1}^{N_{\mathcal{X}}}$ of $\mathcal{X}$ as

$$P_{\mathcal{X}}(t,t') = \sum_{k=1}^{N_{\mathcal{X}}} x_k(t)\, x_k^*(t') . \tag{1.3}$$

The geometry of orthogonal projections is illustrated in **Fig. 1.1**.

Some examples of linear signal spaces frequently encountered in signal processing applications are the space $\mathcal{L}_2(\mathbb{R})$ of all square-integrable (finite-energy) signals, the space of all signals band-limited to a given frequency band, the space of all analytic signals, the space of all signals time-limited to a given time interval, and the space of all causal signals.

1.2 Quadratic Time-Frequency Signal Representations

Besides linear signal spaces, the second basic ingredient of this work is the field of time-frequency (TF) analysis. TF signal representations combine a temporal analysis and a spectral analysis of signals by representing signals over a joint TF plane [Cohen, 1995, Flandrin, 1993, Hlawatsch and Boudreaux-Bartels, 1992, Boashash, 1990, Qian and Chen, 1996]. They are powerful tools for the analysis and processing of "nonstationary" signals. A *quadratic* TF signal representation involves the signal in a quadratic manner. There are two basic modes of interpretation for a quadratic TF signal representation: the "energetic" interpretation as a joint TF energy distribution, and the "correlative" interpretation as a joint TF correlation function [Hlawatsch, 1991]. In this work, we shall concentrate on two specific quadratic TF signal representations, the Wigner distribution and the ambiguity function. We shall briefly review these two TF representations in the following.

- The *Wigner distribution* (WD) [Wigner, 1932, Claasen and Mecklenbräuker, 1980, Claasen and Mecklenbräuker, 1980a, Claasen and Mecklenbräuker, 1980b, Hlawatsch and Flandrin, 1997, Cohen, 1995, Flandrin, 1993, Hlawatsch and Boudreaux-Bartels, 1992, Boashash, 1990] is a prominent "energetic" TF representation. The WD of a (generally complex-valued) signal $x(t)$ with Fourier transform

$$X(f) = \int_t x(t)\, e^{-j2\pi ft}\, dt$$

  is defined as

$$W_x(t,f) = \int_\tau x\!\left(t+\frac{\tau}{2}\right) x^*\!\left(t-\frac{\tau}{2}\right) e^{-j2\pi f\tau}\, d\tau \qquad (1.4)$$

$$= \int_\nu X\!\left(f+\frac{\nu}{2}\right) X^*\!\left(f-\frac{\nu}{2}\right) e^{j2\pi t\nu}\, d\nu\,. \qquad (1.5)$$

  The WD is a real-valued function of time t and frequency f. Its interpretation as a TF energy distribution is primarily based on the *marginal properties*

$$\int_f W_x(t,f)\, df = d_x(t) \qquad (1.6)$$

$$\int_t W_x(t,f)\, dt = D_x(f) \qquad (1.7)$$

$$\int_t \int_f W_x(t,f)\, dt\, df = E_x \qquad (1.8)$$

where the instantaneous power (temporal energy density) $d_x(t)$, the spectral energy density $D_x(f)$, and the energy E_x are defined as

$$d_x(t) = |x(t)|^2$$
$$D_x(f) = |X(f)|^2$$
$$E_x = \int_t |x(t)|^2\, dt = \int_f |X(f)|^2\, df. \tag{1.9}$$

The WD satisfies a large number of further interesting properties [Claasen and Mecklenbräuker, 1980, Hlawatsch and Boudreaux-Bartels, 1992]. An especially important property of the WD is *Moyal's formula* [Claasen and Mecklenbräuker, 1980]

$$\langle W_x, W_y \rangle = |\langle x, y \rangle|^2 \tag{1.10}$$

where $\langle W_x, W_y \rangle = \int_t \int_f W_x(t,f)\, W_y(t,f)\, dt\, df$. Moyal's formula relates the inner product of two WDs to the inner product of the signals involved.

- The (symmetric) *ambiguity function* (AF) [Woodward, 1953, Rihaczek, 1969, Van Trees, 1992, Hlawatsch and Boudreaux-Bartels, 1992, Hlawatsch and Flandrin, 1997, Szu and Blodgett, 1981] is a prominent "correlative" TF representation. The AF of a signal $x(t)$ is defined as

$$A_x(\tau, \nu) = \int_t x\left(t + \frac{\tau}{2}\right) x^*\left(t - \frac{\tau}{2}\right) e^{-j2\pi\nu t}\, dt$$
$$= \int_f X\left(f + \frac{\nu}{2}\right) X^*\left(f - \frac{\nu}{2}\right) e^{j2\pi\tau f}\, df,$$

where τ and ν are, respectively, time lag and frequency lag variables. The "marginal properties" of the AF state that

$$A_x(\tau, 0) = r_x(\tau)$$
$$A_x(0, \nu) = R_x(\nu)$$
$$A_x(0, 0) = E_x.$$

Here, the temporal correlation $r_x(\tau)$ and the spectral correlation $R_x(\nu)$ are defined as

$$r_x(\tau) = \int_t x(t+\tau)\, x^*(t)\, dt$$
$$R_x(\nu) = \int_f X(f+\nu)\, X^*(f)\, df,$$

and $E_x = r_x(0) = R_x(0)$ is the energy of $x(t)$ (cf. (1.9)). The AF is the 2-D Fourier transform of the WD,

$$A_x(\tau, \nu) = \int_t \int_f W_x(t, f)\, e^{-j2\pi(\nu t - \tau f)}\, dt\, df\ .$$

Like the WD, the AF satisfies Moyal's formula,

$$\langle A_x, A_y \rangle = |\langle x, y \rangle|^2\ , \tag{1.11}$$

where $\langle A_x, A_y \rangle = \int_\tau \int_\nu A_x(\tau, \nu)\, A_y^*(\tau, \nu)\, d\tau\, d\nu$.

The results to be presented in subsequent chapters heavily build on the theory of the WD and AF. A detailed review of TF analysis in general and the WD and AF in particular is beyond the scope of this introduction. The interested reader is referred to the pertinent literature [Wigner, 1932, Claasen and Mecklenbräuker, 1980, Claasen and Mecklenbräuker, 1980a, Claasen and Mecklenbräuker, 1980b, Hlawatsch and Flandrin, 1997, Cohen, 1995, Flandrin, 1993, Hlawatsch and Boudreaux-Bartels, 1992, Qian and Chen, 1996, Boashash, 1990, Woodward, 1953, Rihaczek, 1969, Van Trees, 1992].

1.3   Outline and Summary of Results

Following this brief review of linear signal spaces and quadratic TF analysis, we shall now discuss the organization of this work and give a summary of our major results. We recall that the general subject of this work is the extension of the signal representations WD and AF to linear signal spaces, and the discussion of specific signal processing applications that result from this extension.

In **Chapter 2**, the *WD of a linear signal space* is introduced and studied. Similar to the WD of a signal, the WD of a linear signal space describes the space's energy distribution over the TF plane. We derive simple expressions for the WD of a signal space in terms of the space's orthogonal projection operator and orthonormal bases. We study important properties and the "energetic" interpretation of the WD of a signal space, and we consider the results obtained for some specific spaces. The cross-WD of two signal spaces and a discrete-time WD version are introduced. We discuss the minimization and maximization of WD integrals. We also show how any arbitrary quadratic signal representation or quadratic signal parameter can be extended to linear signal spaces.

Chapter 3 uses the WD of a linear signal space for an investigation into the *TF localization of linear signal spaces*. We discuss the geometric properties (or "shape") of the WD of a space. A distinction between "sophisticated" and "simple" spaces is drawn. We discuss the "TF disjointness" and "TF affiliation" of two spaces, and of a signal and a space. Furthermore, two quantities measuring the overall TF concentration of a space are introduced, and concentration bounds are formulated that can be viewed as *uncertainty relations for signal spaces*. The spaces attaining these concentration bounds (i.e., the spaces with maximum TF concentration) are derived. Finally, two related quantities describing the TF concentration and localization of a space in a given TF region are proposed, and bounds for these quantities are derived.

The optimum *TF synthesis* or *TF design* of linear signal spaces is considered in **Chapter 4**. We present a method for constructing a space that, loosely speaking, "comprises all signals located in a given TF region." This can be made mathematically precise using the WD of a space. Specifically, the optimum space is defined as the space whose WD is closest to the indicator function of the given TF region. This optimum space is shown to be an "eigenspace" of the TF region, and some properties of eigenspaces are discussed. The design method is then extended to include a signal subspace constraint, and it is reformulated in a discrete-time setting. Finally, it is shown that the spaces with maximum TF concentration considered in Chapter 3 are also optimum in the context of TF synthesis.

Chapter 5 considers two signal processing applications of the design methods discussed in Chapter 4, namely, *TF projection filters* and *TF signal expansions*. A TF projection filter is a linear, time-varying filter with specified "TF pass region," i.e., the filter passes all signals located in the given TF pass region and suppresses all signals located outside the TF pass region. A TF signal expansion allows the parsimonious representation of signals located in a given TF support region. Both methods permit the effective suppression of noise or other interfering signals. The concept of TF projection filters is extended to *TF filter banks* possessing the perfect reconstruction property. The performance of TF projection filters and TF filter banks is demonstrated using computer simulation.

The application of TF projection filters to statistical signal processing is considered in **Chapter 6**. We derive the optimum projection systems for estimating (enhancing) and detecting a nonstationary random process corrupted by noise. We then show how these statistically optimum projection systems can be approximated by simple TF filter designs that require less statistical *a priori* knowledge than the optimum systems. The satisfactory performance of these approximate TF projection filters is verified using computer simulation. A second type of problems considered is the estimation and detection of signals located in a given TF region.

While Chapters 2 through 6 are dedicated to the WD of a linear signal space and to its various applications, i.e., to the "energetic" type of TF analysis, Chapters 7 and 8 consider the "correlative" TF analysis of linear signal spaces. **Chapter 7** introduces and studies the *AF of a linear signal space*.

Simple expressions in terms of the space's orthogonal projection operator and orthonormal bases are given, important interpretations and properties of the AF of a space are discussed, and the results obtained for some specific spaces are studied. Since the AF and WD of a space are a Fourier transform pair, this chapter is largely analogous to Chapter 2 (discussing the WD of a space). The AF of a space is also shown to allow a simple characterization and interpretation of *sophisticated spaces*.

Chapter 8 considers a potential application of the AF of a linear signal space, namely, the radar/sonar problem of jointly estimating the range and radial velocity of a slowly fluctuating point target. After a brief review of the classical maximum-likelihood *single-pulse* estimator, the maximum-likelihood *multipulse* estimator is derived under the assumption that a number of pulses can be transmitted and received independently of each other, and the Cramér-Rao lower bounds are calculated for this situation. It is shown that a global performance measure of the multipulse estimator is optimized if the transmitted pulses are orthogonal and have equal energies. In this case, the estimator's performance is characterized by the AF of the linear signal space spanned by the transmitted pulses. It is shown that the "thumbtack shape" of the AF of a linear signal space (related to the estimator's performance) can be made arbitrarily good if only the number of pulses transmitted is sufficiently large. This is verified experimentally by studying the AFs of some specific spaces.

Finally, **Chapter 9** contains concluding remarks that summarize and discuss the material presented and indicate possible extensions and suggestions for future research.

2 THE WIGNER DISTRIBUTION OF A LINEAR SIGNAL SPACE

For many good reasons, the WD of a signal is often considered as the "central" quadratic TF signal representation with energetic interpretation. This exceptional role of the WD is maintained when the WD is extended to linear signal spaces, which explains why the WD is being emphasized in this work.

In this chapter, the WD of a linear signal space is introduced and studied, thereby establishing a basis for the material covered in Chapters 3 through 6. We consider the properties and interpretation of the WD of a linear signal space, the results obtained for some specific spaces, and some extensions. Furthermore, the minimization and maximization of TF integrals involving the WD will also be discussed. This optimization problem will repeatedly be encountered in later chapters.

This chapter is organized as follows. Section 2.1 considers two equivalent definitions and several expressions of the WD of a linear signal space. Section 2.2 shows how the principles underlying the definition of the WD of a linear signal space can be extended to arbitrary quadratic signal representations or signal parameters. Some fundamental mathematical properties of the WD of a linear signal space are discussed in Section 2.3. Section 2.4 considers the WDs of several specific signal spaces, and Section 2.5 comments on the "energetic" interpretation of the WD. Two extensions (the cross-WD of two signal spaces and the discrete-time WD) are introduced in Section 2.6. Finally, Section 2.7 considers the minimization and maximization of WD integrals.

2.1 Definitions and Expressions

In this section, we define the *WD of a linear signal space* in two equivalent ways, and we derive simple expressions for the WD of a signal space in terms of the space's orthogonal projection operator and orthonormal bases.

The WD of a linear signal space $\mathcal{X}$ is defined by averaging the WD of a signal (see (1.4)) over all elements $x(t) \in \mathcal{X}$ [Hlawatsch and Kozek, 1993]. This averaging can be performed in a deterministic or stochastic setting. Luckily, we will see that both averaging implementations yield the same result.

2.1.1 Deterministic Definition

The "deterministic" averaging is as follows. Let $\{l_k(t)\}_{k=1}^{\infty}$ be an arbitrary orthonormal basis of $\mathcal{L}_2(\mathbb{R})$, the linear space of all square-integrable (finite-energy) signals. The orthogonal projection operator on $\mathcal{L}_2(\mathbb{R})$ equals the identity operator; its kernel is given by (cf. (1.3))

$$P_{\mathcal{L}_2(\mathbb{R})}(t, t') = \sum_{k=1}^{\infty} l_k(t)\, l_k^*(t') = \delta(t - t')\,.$$

From the second identity, it follows with (1.4) that

$$\sum_{k=1}^{\infty} W_{l_k}(t, f) \equiv 1\,.$$

This shows that the average (sum) of the WDs of all basis signals $l_k(t)$ covers the entire TF plane in an ideally homogeneous manner. Let us now consider the *orthogonal projections* $l_{k,\mathcal{X}}(t)$ of the basis signals $l_k(t)$ onto $\mathcal{X}$ (cf. (1.1), (1.2)). The average (sum) of the WDs of all orthogonal projections $l_{k,\mathcal{X}}(t)$ will be concentrated in those regions of the TF plane where the space's elements $x(t) \in \mathcal{X}$ take on their energy (cf. **Fig. 2.1**). We therefore define the *WD of a linear signal space* $\mathcal{X}$ as

$$W_{\mathcal{X}}(t, f) \triangleq \sum_{k=1}^{\infty} W_{l_{k,\mathcal{X}}}(t, f)\,. \tag{2.1}$$

It can be shown that $W_{\mathcal{X}}(t, f)$ is independent of the specific (orthonormal) basis $\{l_k(t)\}$ of $\mathcal{L}_2(\mathbb{R})$ used in (2.1).

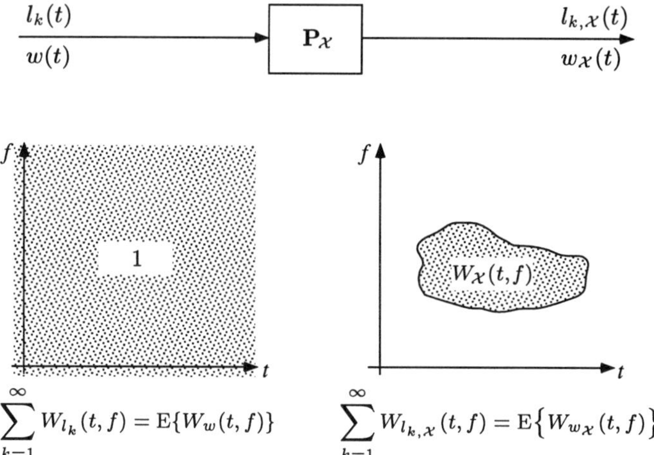

$$\sum_{k=1}^{\infty} W_{l_k}(t,f) = \mathrm{E}\{W_w(t,f)\} \qquad \sum_{k=1}^{\infty} W_{l_{k,\mathcal{X}}}(t,f) = \mathrm{E}\{W_{w_{\mathcal{X}}}(t,f)\}$$

Figure 2.1. Deterministic and stochastic definitions of the WD of a linear signal space.

2.1.2 Stochastic Definition

Let $w(t)$ be wide-sense stationary, zero-mean white noise with normalized power spectral density [Papoulis, 1984a]. The autocorrelation function of $w(t)$ is

$$R_w(t,t') \; \stackrel{\triangle}{=} \; \mathrm{E}\{w(t)\,w^*(t')\} \; = \; \delta(t-t') \qquad (2.2)$$

where E denotes the expectation operator. Combining (2.2) and (1.4), the Wigner-Ville spectrum [Martin and Flandrin, 1985, Flandrin, 1989, Flandrin and Martin, 1997] (i.e., the expected WD) of $w(t)$ is formally obtained as

$$\mathrm{E}\{W_w(t,f)\} \; \equiv \; 1 \, .$$

Hence, the ensemble average (expectation) of the WD of $w(t)$ covers the entire TF plane in an ideally homogeneous manner. In contrast, the ensemble average of the WD of the *orthogonal projection* $w_{\mathcal{X}}(t)$ of $w(t)$ will be concentrated in those TF regions where the space's elements $x(t) \in \mathcal{X}$ take on their energy (cf. Fig. 2.1). We therefore define the *WD of a linear signal space* $\mathcal{X}$ as

$$W_{\mathcal{X}}(t,f) \; \stackrel{\triangle}{=} \; \mathrm{E}\{W_{w_{\mathcal{X}}}(t,f)\} \, . \qquad (2.3)$$

The complete formal analogy of the stochastic and deterministic definitions is obvious. Less obvious, but easily shown, is the strict equivalence of both definitions. Thus, either one of the two definitions (2.1) and (2.3) can be used as the basic definition of the WD of a linear signal space.

2.1.3 Expression in Terms of the Projection Operator

Using (2.1) or (2.3), a straightforward derivation shows that the WD of a signal space $\mathcal{X}$ can be expressed in terms of the space's orthogonal projection operator $\mathbf{P}_{\mathcal{X}}$ as

$$W_{\mathcal{X}}(t,f) = \int_{\tau} (\mathbf{P}_{\mathcal{X}} \mathbf{P}_{\mathcal{X}}^{+})\left(t+\frac{\tau}{2}, t-\frac{\tau}{2}\right) e^{-j2\pi f \tau}\, d\tau \,, \qquad (2.4)$$

where $(\mathbf{P}_{\mathcal{X}} \mathbf{P}_{\mathcal{X}}^{+})(t,t')$ denotes the kernel of the composite operator $\mathbf{P}_{\mathcal{X}} \mathbf{P}_{\mathcal{X}}^{+}$ obtained by cascading the projection operator $\mathbf{P}_{\mathcal{X}}$ and its adjoint $\mathbf{P}_{\mathcal{X}}^{+}$. Since $\mathbf{P}_{\mathcal{X}}$ is self-adjoint and idempotent, $\mathbf{P}_{\mathcal{X}} \mathbf{P}_{\mathcal{X}}^{+} = \mathbf{P}_{\mathcal{X}}$ so that (2.4) reduces to

$$W_{\mathcal{X}}(t,f) = \int_{\tau} P_{\mathcal{X}}\left(t+\frac{\tau}{2}, t-\frac{\tau}{2}\right) e^{-j2\pi f \tau}\, d\tau \,. \qquad (2.5)$$

This simple expression of $W_{\mathcal{X}}(t,f)$ is recognized as the *Weyl symbol* [Kozek, 1992b, Folland, 1989, Janssen, 1989, Shenoy and Parks, 1994, Ramanathan and Topiwala, 1993, Kozek and Hlawatsch, 1991b, Kozek, 1992a] of the projection operator $\mathbf{P}_{\mathcal{X}}$; it is also seen to be reminiscent of the WD of a signal in (1.4). Inversion of (2.5) yields

$$P_{\mathcal{X}}(t_1, t_2) = \int_{f} W_{\mathcal{X}}\left(\frac{t_1 + t_2}{2}, f\right) e^{j2\pi(t_1 - t_2)f}\, df \,,$$

which shows the important fact that the projection operator $\mathbf{P}_{\mathcal{X}}$ can be recovered from the WD $W_{\mathcal{X}}(t,f)$. Hence, the WD of a signal space provides a *complete* characterization of the space.

A "frequency-domain" expression of $W_{\mathcal{X}}(t,f)$ can be derived by noting that the projection $s_{\mathcal{X}}(t)$ in (1.1) can be expressed in the frequency domain as

$$S_{\mathcal{X}}(f) = \int_{f'} \tilde{P}_{\mathcal{X}}(f,f')\, S(f')\, df'$$

where $S(f)$ and $S_{\mathcal{X}}(f)$ are the Fourier transforms of $s(t)$ and $s_{\mathcal{X}}(t)$, respectively, and

$$\tilde{P}_{\mathcal{X}}(f,f') = \int_{t}\int_{t'} P_{\mathcal{X}}(t,t')\, e^{-j2\pi(ft - f't')}\, dt\, dt' \qquad (2.6)$$

is the frequency-domain kernel (bifrequency function) [Zadeh, 1950] of $\mathbf{P}_{\mathcal{X}}$. Inserting (2.6) into (2.5) yields the frequency-domain expression

$$W_{\mathcal{X}}(t,f) = \int_{\nu} \tilde{P}_{\mathcal{X}}\left(f+\frac{\nu}{2}, f-\frac{\nu}{2}\right) e^{j2\pi t\nu}\, d\nu \,,$$

which is analogous to the time-domain expression (2.5) and also to (1.5).

2.1.4 Expression in Terms of a Basis

Inserting (1.3) into (2.5) results in the expression

$$W_{\mathcal{X}}(t, f) \;=\; \sum_{k=1}^{N_{\mathcal{X}}} W_{x_k}(t, f) \tag{2.7}$$

where $\{x_k(t)\}_{k=1}^{N_{\mathcal{X}}}$ is an arbitrary orthonormal basis of $\mathcal{X}$. We see that the WD of a linear signal space is simply the sum of the WDs of all orthonormal basis signals. We emphasize that (2.7) is independent of the specific orthonormal basis of $\mathcal{X}$, as is evidenced by the "basis-free" expression (2.5). Introducing the basis vector $\mathbf{x}(t) = [x_1(t), x_2(t), ..., x_{N_{\mathcal{X}}}(t)]^T$, (2.7) can be written as[1]

$$W_{\mathcal{X}}(t, f) \;=\; \int_{\tau} \mathbf{x}^H\!\left(t - \frac{\tau}{2}\right) \mathbf{x}\!\left(t + \frac{\tau}{2}\right) e^{-j2\pi f\tau}\, d\tau \;,$$

which is again reminiscent of the WD of a signal in (1.4).

Since, according to (2.7), the WD of a linear signal space is the sum of the WDs of several orthonormal signals, we expect the WD of a signal space to cover a larger TF area than the WD of a single signal. This is indeed true in general, except if $N_{\mathcal{X}} = 1$ (see also Chapter 3). **Fig. 2.2** compares the WD of a signal space and the WDs of the orthonormal basis signals spanning the space[2]. **Fig. 2.3** illustrates the fact that the WD of a space is the sum of the WDs of all orthonormal basis signals. (The spaces shown in Figs. 2.2 and 2.3 were synthesized using the TF synthesis methods described in Chapter 4.)

2.2 Quadratic Space Representations

Let us, for a moment, digress from the WD of a signal space to consider an extension of this concept. We shall need this extension in Section 2.3.

2.2.1 Definitions and Expressions

In analogy to the WD of a signal, all other quadratic signal representations can be extended to a linear signal space as well [Hlawatsch and Kozek, 1993]. Any quadratic signal representation $Q_x(\Theta)$ can be written as [Hlawatsch, 1992a]

$$Q_x(\Theta) \;=\; \int_{t_1} \int_{t_2} k_Q(\Theta; t_1, t_2)\, x(t_1)\, x^*(t_2)\, dt_1 dt_2 \;,$$

where $k_Q(\Theta; t_1, t_2)$ is a kernel function specifying the representation Q and Θ is a parameter or parameter vector (e.g., $\Theta = (t, f)$ in the case of the WD). To a

---

[1]The superscripts ^{T} and ^{H} denote transposition and conjugate transposition, respectively.
[2]The contour-line plots in Fig. 2.2 and subsequent figures show positive heights only.

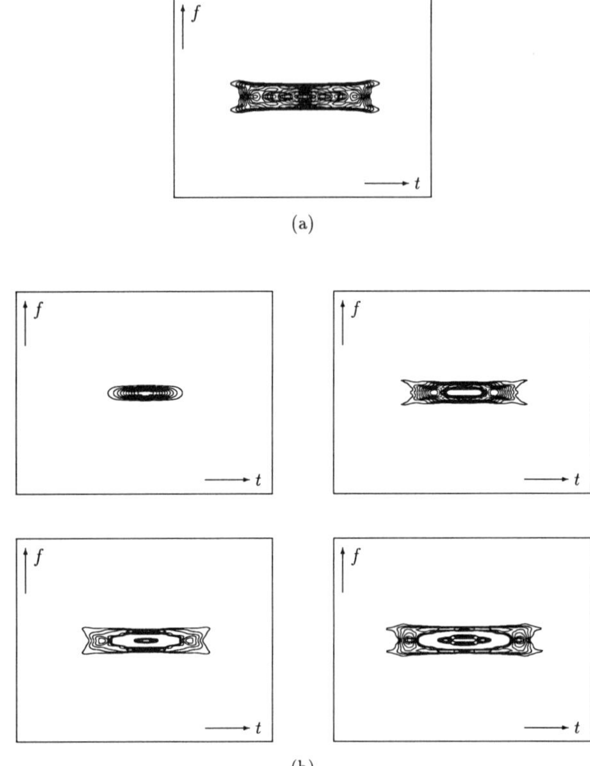

Figure 2.2. (a) WD of a linear signal space with dimension $N_\mathcal{X} = 4$, (b) WDs of the individual orthonormal basis signals.

given quadratic *signal* representation $Q_x(\Theta)$, we define the corresponding *space* representation $Q_\mathcal{X}(\Theta)$ by generalization of the WD definition (2.1) or (2.3),

$$Q_\mathcal{X}(\Theta) \overset{\triangle}{=} \sum_{k=1}^{\infty} Q_{l_k,\mathcal{X}}(\Theta) = \mathrm{E}\{Q_{w_\mathcal{X}}(\Theta)\}. \tag{2.8}$$

From (2.8), expressions in terms of the orthogonal projection operator $\mathbf{P}_\mathcal{X}$ or an orthonormal basis $\{x_k(t)\}_{k=1}^{N_\mathcal{X}}$ are obtained as

$$Q_\mathcal{X}(\Theta) = \int_{t_1} \int_{t_2} k_Q(\Theta; t_1, t_2)\, P_\mathcal{X}(t_1, t_2)\, dt_1 dt_2 = \sum_{k=1}^{N_\mathcal{X}} Q_{x_k}(\Theta).$$

Note that these expressions are generalizations of (2.5) and (2.7).

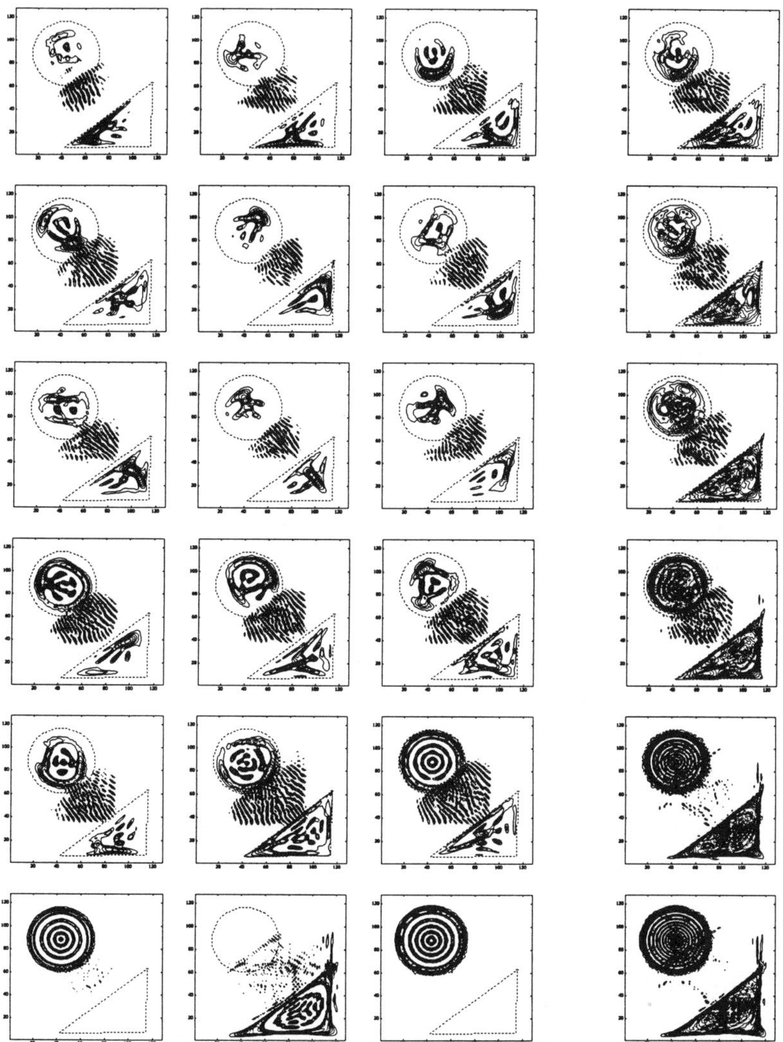

Figure 2.3. The WD of a linear signal space equals the sum of the WDs of all orthonormal basis signals. This is illustrated here for a space $\mathcal{X}$ with dimension $N_{\mathcal{X}} = 18$. The left-hand side shows the individual WDs $W_{x_k}(t, f)$ of the 18 orthonormal basis signals $x_k(t)$, while the right-hand side shows the sums of these WDs, $\sum_{k=1}^{K} W_{x_k}(t, f)$, up to (and including) the respective row. In particular, the last plot shows the WD of the overall space, $W_{\mathcal{X}}(t, f) = \sum_{k=1}^{N_{\mathcal{X}}} W_{x_k}(t, f)$.

2.2.2 Examples

It is obvious that the WD of a linear signal space is a special case of the general definition (2.8). We shall now list some further important examples of quadratic signal representations and the corresponding space representations.

Energy:

$$E_x = \int_t |x(t)|^2 \, dt \quad \Rightarrow \quad E_\mathcal{X} = \int_t P_\mathcal{X}(t,t) \, dt = N_\mathcal{X} \,.$$

Temporal energy density (instantaneous power):

$$d_x(t) = |x(t)|^2 \quad \Rightarrow \quad d_\mathcal{X}(t) = P_\mathcal{X}(t,t) \,.$$

Spectral energy density:

$$D_x(f) = |X(f)|^2 \quad \Rightarrow \quad D_\mathcal{X}(f) = \tilde{P}_\mathcal{X}(f,f) \,.$$

Temporal correlation:

$$r_x(\tau) = \int_t x(t+\tau)\, x^*(t) \, dt \quad \Rightarrow \quad r_\mathcal{X}(\tau) = \int_t P_\mathcal{X}(t+\tau,t) \, dt \,.$$

Spectral correlation:

$$R_x(\nu) = \int_f X(f+\nu)\, X^*(f) \, df \quad \Rightarrow \quad R_\mathcal{X}(\nu) = \int_f \tilde{P}_\mathcal{X}(f+\nu,f) \, df \,.$$

Temporal nth-order moment:

$$m_x^{(n)} = \int_t t^n \, |x(t)|^2 \, dt \quad \Rightarrow \quad m_\mathcal{X}^{(n)} = \int_t t^n P_\mathcal{X}(t,t) \, dt \,. \tag{2.9}$$

Spectral nth-order moment:

$$M_x^{(n)} = \int_f f^n \, |X(f)|^2 \, df \quad \Rightarrow \quad M_\mathcal{X}^{(n)} = \int_f f^n \tilde{P}_\mathcal{X}(f,f) \, df \,. \tag{2.10}$$

Ambiguity function:

$$A_x(\tau,\nu) = \int_t x\!\left(t+\frac{\tau}{2}\right) x^*\!\left(t-\frac{\tau}{2}\right) e^{-j2\pi\nu t} \, dt \quad \Rightarrow$$

$$A_\mathcal{X}(\tau,\nu) = \int_t P_\mathcal{X}\!\left(t+\frac{\tau}{2}, t-\frac{\tau}{2}\right) e^{-j2\pi\nu t} \, dt \,.$$

Spectrogram using analysis window $h(t)$ [Claasen and Mecklenbräuker, 1980b, Flandrin, 1993, Hlawatsch and Boudreaux-Bartels, 1992, Hlawatsch and Flandrin, 1997, Altes, 1980, Kadambe and Boudreaux-Bartels, 1992]:

$$S_x^{(h)}(t,f) = \left| \int_{t'} x(t')\, h^*(t'-t)\, e^{-j2\pi ft'} dt' \right|^2 \quad \Rightarrow$$

$$S_{\mathcal{X}}^{(h)}(t,f) = \int_{t_1}\int_{t_2} P_{\mathcal{X}}(t_1,t_2)\, h^*(t_1-t)\, h(t_2-t)\, e^{-j2\pi f(t_1-t_2)}\, dt_1 dt_2 \geq 0. \quad (2.11)$$

Scalogram using analysis wavelet $h(t)$ with center frequency f_r [Flandrin, 1993, Rioul and Flandrin, 1992, Hlawatsch and Boudreaux-Bartels, 1992]:

$$C_x^{(h)}(t,f) = \left| \int_{t'} x(t')\, \sqrt{\left|\frac{f}{f_r}\right|}\, h^*\left(\frac{f}{f_r}(t'-t)\right) dt' \right|^2 \quad \Rightarrow$$

$$C_{\mathcal{X}}^{(h)}(t,f) =$$

$$\int_{t_1}\int_{t_2} P_{\mathcal{X}}(t_1,t_2)\, \left|\frac{f}{f_r}\right|\, h^*\left(\frac{f}{f_r}(t_1-t)\right) h\left(\frac{f}{f_r}(t_2-t)\right) dt_1 dt_2 \geq 0. \quad (2.12)$$

Note that the energy $E_{\mathcal{X}}$ of a linear signal space equals the space's dimension $N_{\mathcal{X}}$. The spectrogram and scalogram will be considered in Section 2.5, and the ambiguity function will be discussed in Chapters 7 and 8.

Quite generally, the following statement can be shown: If two quadratic signal representations $Q_x(\Theta)$ and $Q'_x(\Theta')$ are related via a linear transformation, then the corresponding space representations $Q_{\mathcal{X}}(\Theta)$ and $Q'_{\mathcal{X}}(\Theta')$ are related via the *same* linear transformation. This general principle will be specialized to the WD in the next section.

2.3 Properties

After this digression to general "quadratic space representations," we return to our current topic, the WD of a linear signal space. In this section, some properties of the WD of a space [Hlawatsch and Kozek, 1993] (most of which are analogous to well-known properties of the WD of a signal [Claasen and Mecklenbräuker, 1980, Hlawatsch and Boudreaux-Bartels, 1992, Flandrin, 1993]) are stated without proof.

1) Real-valued. The WD of a space is a real-valued function which, however, is not guaranteed to be everywhere nonnegative.

2) Time-frequency integral. The integral of $W_{\mathcal{X}}(t,f)$ over the entire TF plane equals the dimension of $\mathcal{X}$,

$$\int_t \int_f W_{\mathcal{X}}(t,f)\, dt\, df = N_{\mathcal{X}}.$$

3) Norm. The squared norm of $W_{\mathcal{X}}(t,f)$ equals the dimension of $\mathcal{X}$,

$$\|W_{\mathcal{X}}\|^2 = \int_t \int_f \left[W_{\mathcal{X}}(t,f) \right]^2 dt\, df = N_{\mathcal{X}} \,. \tag{2.13}$$

This shows that $W_{\mathcal{X}}(t,f)$ is square-integrable if and only if $\mathcal{X}$ has finite dimension. Combining properties 2) and 3) gives the identity

$$\int_t \int_f \left[W_{\mathcal{X}}(t,f) \right]^2 dt\, df = \int_t \int_f W_{\mathcal{X}}(t,f)\, dt\, df \tag{2.14}$$

which restricts the behavior of the function $W_{\mathcal{X}}(t,f)$. Clearly, (2.14) will be satisfied if $W_{\mathcal{X}}(t,f)$ assumes only values 0 or 1. Such a "binary" WD is indeed obtained for some specific spaces with infinite dimension (see Section 2.4). More generally, in many cases $W_{\mathcal{X}}(t,f)$ tends to be oscillatory around the heights 0 and 1. The shape of $W_{\mathcal{X}}(t,f)$ will be further discussed in Section 3.1.

4) Finite support. If all signals $x(t) \in \mathcal{X}$ are time-limited to an interval $[t_1, t_2]$ (or, equivalently, if $\mathcal{X}$ is a subspace of the space $\mathcal{T}[t_1, t_2]$ of all signals time-limited to $[t_1, t_2]$), then $W_{\mathcal{X}}(t,f)$ is zero for all t outside $[t_1, t_2]$,

$$\mathcal{X} \subseteq \mathcal{T}[t_1, t_2] \quad \Rightarrow \quad W_{\mathcal{X}}(t,f) = 0 \quad \text{for } t \notin [t_1, t_2] \,.$$

Similarly, if all signals $x(t) \in \mathcal{X}$ are band-limited to a frequency band $[f_1, f_2]$ (i.e., if $\mathcal{X}$ is a subspace of the space $\mathcal{F}[f_1, f_2]$ of all signals band-limited to $[f_1, f_2]$), then $W_{\mathcal{X}}(t,f)$ is zero for all f outside $[f_1, f_2]$,

$$\mathcal{X} \subseteq \mathcal{F}[f_1, f_2] \quad \Rightarrow \quad W_{\mathcal{X}}(t,f) = 0 \quad \text{for } f \notin [f_1, f_2] \,. \tag{2.15}$$

These properties are analogous to the "finite-support properties" satisfied by the WD of a signal [Claasen and Mecklenbräuker, 1980].

5) Moyal-type relation I. The inner product of the WDs of two signal spaces $\mathcal{X}$ and $\mathcal{Y}$,

$$\langle W_{\mathcal{X}}, W_{\mathcal{Y}} \rangle = \int_t \int_f W_{\mathcal{X}}(t,f)\, W_{\mathcal{Y}}(t,f)\, dt\, df \,,$$

can be expressed in terms of the spaces' orthogonal projection operators $\mathbf{P}_{\mathcal{X}}$ and $\mathbf{P}_{\mathcal{Y}}$ or the spaces' orthonormal bases $\{x_k(t)\}_{k=1}^{N_{\mathcal{X}}}$ and $\{y_l(t)\}_{l=1}^{N_{\mathcal{Y}}}$ as follows:

$$\langle W_{\mathcal{X}}, W_{\mathcal{Y}} \rangle = \int_t \int_{t'} P_{\mathcal{X}}(t,t')\, P_{\mathcal{Y}}^*(t,t')\, dt\, dt' = \sum_{k=1}^{N_{\mathcal{X}}} \sum_{l=1}^{N_{\mathcal{Y}}} |\langle x_k, y_l \rangle|^2 \,. \tag{2.16}$$

This can be viewed as a generalization of Moyal's formula (1.10).

Using (2.16), it can be shown that $\langle W_{\mathcal{X}}, W_{\mathcal{Y}} \rangle$ is bounded as

$$0 \leq \langle W_{\mathcal{X}}, W_{\mathcal{Y}} \rangle \leq \min\{N_{\mathcal{X}}, N_{\mathcal{Y}}\} \left(\leq \sqrt{N_{\mathcal{X}} N_{\mathcal{Y}}} \right). \tag{2.17}$$

The lower bound is attained if and only if the two spaces are orthogonal,

$$\langle W_{\mathcal{X}}, W_{\mathcal{Y}} \rangle = 0 \qquad \Longleftrightarrow \qquad \mathcal{X} \perp \mathcal{Y}. \tag{2.18}$$

The upper bound is attained if and only if one space is a subspace of the other space. Assuming, for example, $N_{\mathcal{Y}} \leq N_{\mathcal{X}}$ so that $\min\{N_{\mathcal{X}}, N_{\mathcal{Y}}\} = N_{\mathcal{Y}}$, we have

$$\langle W_{\mathcal{X}}, W_{\mathcal{Y}} \rangle = N_{\mathcal{Y}} \qquad \Longleftrightarrow \qquad \mathcal{Y} \subseteq \mathcal{X}. \tag{2.19}$$

Geometric implications of these results will be discussed in Section 3.2.

6) Moyal-type relation II. The inner product of the WD of a signal $s(t)$ and the WD of a signal space $\mathcal{X}$ equals the energy of the projected signal,

$$\langle W_s, W_{\mathcal{X}} \rangle = \|s_{\mathcal{X}}\|^2 = \sum_{k=1}^{N_{\mathcal{X}}} |\langle s, x_k \rangle|^2. \tag{2.20}$$

This can again be considered a generalization of Moyal's formula (1.10). From (2.20), it follows that $\langle W_s, W_{\mathcal{X}} \rangle$ is bounded as

$$0 \leq \langle W_s, W_{\mathcal{X}} \rangle \leq \|s\|^2. \tag{2.21}$$

The lower bound is attained if and only if $s(t)$ is orthogonal to $\mathcal{X}$,

$$\langle W_s, W_{\mathcal{X}} \rangle = 0 \qquad \Longleftrightarrow \qquad s \perp \mathcal{X}, \tag{2.22}$$

whereas the upper bound is attained if and only if $s(t)$ is an element of $\mathcal{X}$,

$$\langle W_s, W_{\mathcal{X}} \rangle = \|s\|^2 \qquad \Longleftrightarrow \qquad s \in \mathcal{X}. \tag{2.23}$$

Again, geometric implications of these results will be discussed in Section 3.2.

7) Direct sum. If two spaces $\mathcal{X}$ and $\mathcal{Y}$ are orthogonal, then the WD of their (direct) sum [Naylor and Sell, 1982, Luenberger, 1969] $\mathcal{X} + \mathcal{Y} = \{x + y : x \in \mathcal{X}, y \in \mathcal{Y}\}$ equals the sum of the WDs of $\mathcal{X}$ and $\mathcal{Y}$,

$$\mathcal{X} \perp \mathcal{Y} \qquad \Rightarrow \qquad W_{\mathcal{X}+\mathcal{Y}}(t, f) = W_{\mathcal{X}}(t, f) + W_{\mathcal{Y}}(t, f). \tag{2.24}$$

In particular, with $\bar{\mathcal{X}}$ denoting the orthogonal complement of $\mathcal{X}$ (i.e., $\mathcal{X} \perp \bar{\mathcal{X}}$

and $\mathcal{X} + \bar{\mathcal{X}} = \mathcal{L}_2(\mathbb{R})$), it follows from (2.24) and from $W_{\mathcal{L}_2(\mathbb{R})}(t, f) \equiv 1$ (see (2.38)) that

$$W_{\bar{\mathcal{X}}}(t, f) = 1 - W_{\mathcal{X}}(t, f) .$$

For non-orthogonal spaces $\mathcal{X}$ and $\mathcal{Y}$, the WD of $\mathcal{X} + \mathcal{Y}$ cannot be expressed in terms of $W_{\mathcal{X}}(t, f)$ and $W_{\mathcal{Y}}(t, f)$ in a simple way.

8) Derivation of other quadratic space representations. Any quadratic space representation (cf. Section 2.2)

$$Q_{\mathcal{X}}(\Theta) = \int_{t_1} \int_{t_2} k_Q(\Theta; t_1, t_2) \, P_{\mathcal{X}}(t_1, t_2) \, dt_1 dt_2$$

can be derived from the WD $W_{\mathcal{X}}(t, f)$ via a linear transformation,

$$Q_{\mathcal{X}}(\Theta) = \int_t \int_f L_Q(\Theta; t, f) \, W_{\mathcal{X}}(t, f) \, dt \, df ,$$

where the transformation kernel $L_Q(\Theta; t, f)$ is given by

$$L_Q(\Theta; t, f) = \int_\tau k_Q\left(\Theta; t + \frac{\tau}{2}, t - \frac{\tau}{2}\right) e^{j2\pi f\tau} \, d\tau .$$

We emphasize that this transformation relating $W_{\mathcal{X}}(t, f)$ and $Q_{\mathcal{X}}(\Theta)$ is the same as the transformation relating the corresponding *signal* representations $W_x(t, f)$ and $Q_x(\Theta)$ [Hlawatsch, 1992a]. Therefore, any linear relation connecting the WD of a signal with some other quadratic signal representation can immediately be reformulated for a space. Applying this principle to the quadratic signal/space representations considered in Section 2.2.2, we obtain the following relations:

$$E_{\mathcal{X}} = \int_t \int_f W_{\mathcal{X}}(t, f) \, dt \, df \tag{2.25}$$

$$d_{\mathcal{X}}(t) = \int_f W_{\mathcal{X}}(t, f) \, df \tag{2.26}$$

$$D_{\mathcal{X}}(f) = \int_t W_{\mathcal{X}}(t, f) \, dt \tag{2.27}$$

$$r_{\mathcal{X}}(\tau) = \int_t \int_f W_{\mathcal{X}}(t, f) \, e^{j2\pi\tau f} \, dt \, df$$

$$R_{\mathcal{X}}(\nu) = \int_t \int_f W_{\mathcal{X}}(t, f) \, e^{-j2\pi\nu t} \, dt \, df$$

$$m_{\mathcal{X}}^{(n)} = \int_t \int_f t^n \, W_{\mathcal{X}}(t, f) \, dt \, df$$

$$M_{\mathcal{X}}^{(n)} = \int_t \int_f f^n \, W_{\mathcal{X}}(t,f) \, dt \, df$$

$$A_{\mathcal{X}}(\tau,\nu) = \int_t \int_f W_{\mathcal{X}}(t,f) \, e^{-j2\pi(\nu t - \tau f)} \, dt \, df \tag{2.28}$$

$$S_{\mathcal{X}}^{(h)}(t,f) = \int_{t'} \int_{f'} W_h(t'-t, f'-f) \, W_{\mathcal{X}}(t',f') \, dt' \, df' \tag{2.29}$$

$$C_{\mathcal{X}}^{(h)}(t,f) = \int_{t'} \int_{f'} W_h\left(\frac{f}{f_r}(t'-t), \frac{f_r}{f}f'\right) W_{\mathcal{X}}(t',f') \, dt' \, df' . \tag{2.30}$$

In particular, the first three relations can be interpreted as "marginal properties" stating that the energy $E_{\mathcal{X}}$ and the energy densities $d_{\mathcal{X}}(t)$ and $D_{\mathcal{X}}(f)$ can be derived from the WD $W_{\mathcal{X}}(t,f)$ by integrating with respect to time and/or frequency (cf. (1.6)–(1.8)).

9) Linear space transformations—General. Let $\mathbf{H}$ denote a linear signal transformation, i.e., a linear, generally time-varying system with input-output relation

$$(\mathbf{H}x)(t) = \int_{t'} H(t,t') \, x(t') \, dt' \, ,$$

where $H(t,t')$ is the kernel (impulse response) of $\mathbf{H}$. The collection of all signals $(\mathbf{H}x)(t)$ with $x(t) \in \mathcal{X}$ is a linear signal space that will be denoted as $\mathbf{H}\mathcal{X}$. If the operator $\mathbf{H}$ is unitary on $\mathcal{L}_2(\mathbb{R})$, i.e., if $\mathbf{H}$ maps $\mathcal{L}_2(\mathbb{R})$ onto $\mathcal{L}_2(\mathbb{R})$ and $\langle \mathbf{H}x, \mathbf{H}y \rangle = \langle x, y \rangle$ for all $x, y \in \mathcal{L}_2(\mathbb{R})$, then the WD of the transformed space $\mathbf{H}\mathcal{X}$ can be derived from the WD of $\mathcal{X}$ via a linear transformation,

$$W_{\mathbf{H}\mathcal{X}}(t,f) = \int_{t'} \int_{f'} L_H(t,f;t',f') \, W_{\mathcal{X}}(t',f') \, dt' \, df' , \tag{2.31}$$

where the kernel $L_H(t,f;t',f')$ is given by

$$L_H(t,f;t',f') = \int_\tau \int_{\tau'} H\left(t+\frac{\tau}{2}, t'+\frac{\tau'}{2}\right) H^*\left(t-\frac{\tau}{2}, t'-\frac{\tau'}{2}\right) e^{-j2\pi(f\tau - f'\tau')} \, d\tau \, d\tau' . \tag{2.32}$$

This linear transformation relating the WDs of the spaces $\mathcal{X}$ and $\mathbf{H}\mathcal{X}$ is identical to the linear transformation relating the WDs of the signals $x(t)$ and $(\mathbf{H}x)(t)$ [Hlawatsch, 1992a], the only difference being that (2.31), (2.32) hold in the "signal case" even if the operator $\mathbf{H}$ is non-unitary. Therefore, if the unitarity condition is met, then a relation known to hold in the signal case can immediately be reformulated for spaces. The next two sets of properties are examples of this principle.

10) Time-frequency coordinate transforms. An important class of unitary signal transformations corresponds to affine, area-preserving TF coordinate transforms [Janssen, 1982]

$$(t, f) \longrightarrow (\alpha t + \beta f - \tau, \gamma t + \delta f - \nu) \qquad \text{with} \quad D = \alpha\delta - \gamma\beta = 1 .$$

To any such TF coordinate transform, a linear operator $\mathbf{H}$ can be associated such that

$$W_{\mathbf{H}\mathcal{X}}(t, f) = W_{\mathcal{X}}(\alpha t + \beta f - \tau, \gamma t + \delta f - \nu) .$$

Some special cases of particular importance are listed below.

TF shift:

$$(\mathbf{H}x)(t) = x(t-\tau)\, e^{j2\pi\nu t} \quad \Rightarrow \quad W_{\mathbf{H}\mathcal{X}}(t, f) = W_{\mathcal{X}}(t-\tau, f-\nu) . \tag{2.33}$$

TF scaling:

$$(\mathbf{H}x)(t) = \sqrt{|a|}\, x(at) \quad \Rightarrow \quad W_{\mathbf{H}\mathcal{X}}(t, f) = W_{\mathcal{X}}\left(at, \frac{f}{a}\right) .$$

Chirp convolution:

$$(\mathbf{H}x)(t) = x(t) * \sqrt{|c|}\, e^{j\pi ct^2} \quad \Rightarrow \quad W_{\mathbf{H}\mathcal{X}}(t, f) = W_{\mathcal{X}}\left(t - \frac{f}{c}, f\right) . \tag{2.34}$$

Chirp multiplication:

$$(\mathbf{H}x)(t) = x(t)\, e^{j\pi ct^2} \quad \Rightarrow \quad W_{\mathbf{H}\mathcal{X}}(t, f) = W_{\mathcal{X}}(t, f-ct) . \tag{2.35}$$

Fourier transform:

$$(\mathbf{H}x)(t) = \sqrt{|c|}\, X(ct) \quad \Rightarrow \quad W_{\mathbf{H}\mathcal{X}}(t, f) = W_{\mathcal{X}}\left(-\frac{f}{c}, ct\right) .$$

These relations show that the WD of a space, just like the WD of a signal, is "covariant" to TF shifts and scalings, convolutions and multiplications by chirp signals, and Fourier transforms. "Covariance" means that these space transformations do not change the space's WD apart from a specific TF coordinate transform. An example is depicted in **Fig. 2.4**.

11) Convolution and multiplication. Another class of unitary transformations corresponds to a linear, time-invariant allpass filter with impulse response $h(t)$ and frequency response $H(f)$ with $|H(f)| \equiv 1$. We here obtain

$$(\mathbf{H}x)(t) = \int_{t'} h(t - t')\, x(t')\, dt' \quad \text{with} \quad |H(f)| \equiv 1$$

$$\Rightarrow \quad W_{\mathbf{H}\mathcal{X}}(t, f) = \int_{t'} W_h(t - t', f)\, W_{\mathcal{X}}(t', f)\, dt' . \tag{2.36}$$

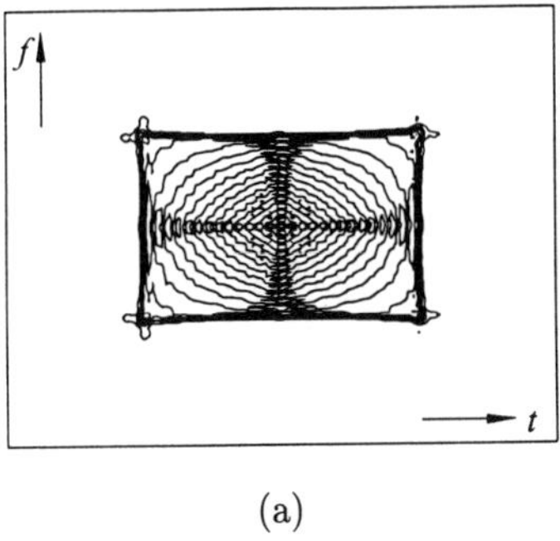
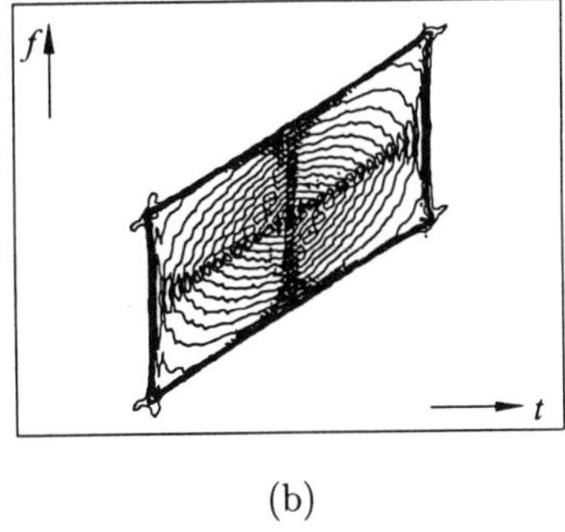

(a)   (b)

Figure 2.4. Multiplication of a linear signal space by a chirp signal: (a) WD of original signal space, (b) WD of transformed signal space. The multiplication of the space by the chirp signal produces a shearing of the space's WD as described by Eq. (2.35).

Similarly, for the multiplication by a signal $m(t)$ with constant (unity) envelope $|m(t)| \equiv 1$ we find

$$(\mathbf{H}x)(t) = m(t)\,x(t) \quad \text{with } |m(t)| \equiv 1$$

$$\Rightarrow \quad W_{\mathbf{H}\mathcal{X}}(t,f) = \int_{f'} W_m(t, f - f')\,W_{\mathcal{X}}(t, f')\,df'. \qquad (2.37)$$

The conditions $|H(f)| \equiv 1$ and $|m(t)| \equiv 1$ guarantee the unitarity of the corresponding transformation $\mathbf{H}$. Note that (2.34) and (2.35) are special cases of (2.36) and (2.37), respectively.

2.4 Some Signal Spaces

It is instructive to consider the WD of a few specific signal spaces [Hlawatsch and Kozek, 1993]. For some spaces, simple WD expressions are obtained. In other cases, the WD will be studied by means of computer simulation.

1) Zero space. The WD of the trivial "zero space" $\mathcal{Z}$ whose only element is the zero signal $z(t) \equiv 0$ is identically zero,

$$W_{\mathcal{Z}}(t, f) \equiv 0\,.$$

2) One-dimensional space. The WD of a one-dimensional space $\mathcal{X}$ equals the WD of the single normalized basis signal $x_1(t)$,

$$W_{\mathcal{X}}(t, f) = W_{x_1}(t, f) \qquad \text{for } N_{\mathcal{X}} = 1\,.$$

Thus, the WD of a (normalized) signal is a special case of the WD of a space.

3) Space of finite-energy signals. The WD of the infinite-dimensional "total space" $\mathcal{L}_2(\mathbb{R})$ of all finite-energy signals is identically unity,

$$W_{\mathcal{L}_2(\mathbb{R})}(t, f) \equiv 1 . \tag{2.38}$$

Here, the entire TF plane is uniformly and homogeneously covered with energy.

4) Space of even signals. The WD of the space $\mathcal{E}$ comprising all even signals (i.e., signals $x(t)$ satisfying $x(-t) = x(t)$) is[3]

$$W_{\mathcal{E}}(t, f) = \frac{1}{2} + \frac{1}{4}\,\delta(t)\,\delta(f) . \tag{2.39}$$

5) Space of odd signals. The WD of the space $\mathcal{O}$ comprising all odd signals (i.e., signals satisfying $x(-t) = -x(t)$) is

$$W_{\mathcal{O}}(t, f) = \frac{1}{2} - \frac{1}{4}\,\delta(t)\,\delta(f) . \tag{2.40}$$

6) Space of time-limited signals. The WD of the space $\mathcal{T}[t_1, t_2]$ of all signals that are time-limited to a given interval $[t_1, t_2]$ equals unity for t inside $[t_1, t_2]$ and zero for t outside $[t_1, t_2]$,

$$W_{\mathcal{T}[t_1,t_2]}(t, f) = \begin{cases} 1, & t \in [t_1, t_2] \\ 0, & t \notin [t_1, t_2] . \end{cases}$$

7) Space of causal signals. Letting $t_1 = 0$ and $t_2 = \infty$ yields the space $\mathcal{C} = \mathcal{T}[0, \infty)$ of causal signals, for which

$$W_{\mathcal{C}}(t, f) = \begin{cases} 1, & t \geq 0 \\ 0, & t < 0 . \end{cases} \tag{2.41}$$

8) Space of band-limited signals. The WD of the space $\mathcal{F}[f_1, f_2]$ of all signals that are band-limited to a given frequency band $[f_1, f_2]$ is obtained as

$$W_{\mathcal{F}[f_1,f_2]}(t, f) = \begin{cases} 1, & f \in [f_1, f_2] \\ 0, & f \notin [f_1, f_2] . \end{cases}$$

---

[3]In the context of complex-valued signals, it is more usual to consider *conjugate even* signals, i.e., signals with $x^*(-t) = x(t)$. However, these signals do not form a linear space. An analogous statement holds for conjugate odd signals.

9) Space of analytic signals. Letting $f_1 = 0$ and $f_2 = \infty$ yields the space $\mathcal{A} = \mathcal{F}[0, \infty)$ of analytic signals (also known as *Hardy space*), for which

$$W_{\mathcal{A}}(t, f) = \begin{cases} 1, & f \geq 0 \\ 0, & f < 0 . \end{cases} \qquad (2.42)$$

10) Hermite spaces. We define the N-dimensional *Hermite space*

$$\mathcal{H}_N^{(T)} \triangleq \text{span}\{h_k^{(T)}(t)\}_{k=1}^N$$

as the space spanned by the first N (orthonormal) Hermite functions [Janssen, 1997b, Folland, 1989, de Bruijn, 1967, Klauder, 1960, Wilcox, 1991]

$$h_k^{(T)}(t) = C_{k-1} \frac{1}{\sqrt{T}} H_{k-1}\left(\sqrt{2\pi}\, \frac{t}{T}\right) e^{-\pi(t/T)^2}, \qquad k = 1, 2, ..., N , \quad (2.43)$$

where

$$H_n(\alpha) = (-1)^n e^{\alpha^2} \frac{d^n}{d\alpha^n} e^{-\alpha^2}, \qquad n = 0, 1, ...$$

is the Hermite polynomial [Abramowitz and Stegun, 1965] of order n, $T > 0$ is a time scaling parameter, and $C_n = 1/\sqrt{2^{n-1/2}\, n!}$.

The WD of the kth Hermite function $h_k^{(T)}(t)$ is given by [Janssen, 1997b]

$$W_{h_k^{(T)}}(t, f) = v_{k-1}\left(\left(\frac{t}{T}\right)^2 + (Tf)^2\right), \qquad k = 1, 2, ... \qquad (2.44)$$

with

$$v_n(\beta) = 2(-1)^n L_n(4\pi\beta)\, e^{-2\pi\beta}, \qquad n = 0, 1, ... , \qquad (2.45)$$

where

$$L_n(\beta) = \frac{1}{n!} e^\beta \frac{d^n}{d\beta^n}\left(\beta^n e^{-\beta}\right) = \sum_{i=0}^n \binom{n}{i} \frac{(-\beta)^i}{i!}, \quad \beta \geq 0, \ n = 0, 1, ...$$

$$(2.46)$$

is the Laguerre polynomial [Abramowitz and Stegun, 1965] of order n. With $W_{\mathcal{H}_N^{(T)}}(t, f) = \sum_{k=1}^N W_{h_k^{(T)}}(t, f)$, it follows that

$$W_{\mathcal{H}_N^{(T)}}(t, f) = w_N\left(\left(\frac{t}{T}\right)^2 + (Tf)^2\right)$$

with

$$w_N(\beta) \;=\; \sum_{n=0}^{N-1} v_n(\beta) \;=\; 2\,e^{-2\pi\beta} \sum_{n=0}^{N-1} (-1)^n\, L_n(4\pi\beta)\,.$$

We see that the WD of $\mathcal{H}_N^{(T)}$ shows "elliptical symmetry," i.e., its height is constant along any ellipse $(t/T)^2 + (Tf)^2 = K^2$ in the TF plane. The shape of the WD in any radial direction is determined by the function $w_N(\beta)$ that has been shown in [Duijvelaar, 1984] to be oscillatory about 1 for $\beta < N/\pi$ (except for a characteristic irregularity at $\beta = 0$), and for $\beta > N/\pi$ to be exponentially decaying to zero as $\beta \to \infty$. From these results, it follows that $W_{\mathcal{H}_N^{(T)}}(t, f)$ is highly concentrated, and oscillatory about height 1, inside a TF region bounded by the ellipse $(t/T_N)^2 + (f/F_N)^2 = 1$ with $T_N = \sqrt{N/\pi}\, T$ and $F_N = \sqrt{N/\pi}\, \frac{1}{T}$. The area enclosed by this ellipse is N. Thus, the area of the effective TF support of the space $\mathcal{H}_N^{(T)}$ is equal to the space's dimension N, and the effective time support and frequency support of $\mathcal{H}_N^{(T)}$ is $[-T_N, T_N]$ and $[-F_N, F_N]$, respectively.

For $N = 1$, the Hermite space reduces to the single Gaussian function $h_1^{(T)}(t) = \sqrt{\sqrt{2}/T}\, e^{-\pi(t/T)^2}$, and its WD is a two-dimensional Gaussian,

$$W_{\mathcal{H}_1^{(T)}}(t, f) \;=\; W_{h_1^{(T)}}(t, f) \;=\; 2\,e^{-2\pi\left[(t/T)^2 + (Tf)^2\right]}\,.$$

For $N \to \infty$, on the other hand, $\mathcal{H}_N^{(T)}$ becomes $\mathcal{L}_2(\mathbb{R})$, so that

$$W_{\mathcal{H}_\infty^{(T)}}(t, f) \;=\; W_{\mathcal{L}_2(\mathbb{R})}(t, f) \;\equiv\; 1\,.$$

Fig. 2.5 shows the WD of the Hermite space $\mathcal{H}_N^{(T)}$ with dimension $N = 25$. The WDs of $\mathcal{H}_N^{(T)}$ for various dimensions N are compared in **Fig. 2.6** (note that, due to the elliptical symmetry of $W_{\mathcal{H}_N^{(T)}}(t, f)$, a cross section through the origin of the TF plane suffices to describe the overall shape of $W_{\mathcal{H}_N^{(T)}}(t, f)$).

The Hermite spaces $\mathcal{H}_N^{(T)}$ will be further considered in Sections 3.3.2, 4.6.1, 7.3, and 8.3.2. In particular, they will be shown in Sections 3.3.2 and 4.6.1 to have maximum TF concentration.

11) Prolate spheroidal spaces. The Hermite spaces are highly concentrated with respect to both time and frequency, but they are neither strictly time-limited nor strictly band-limited. In contrast, the "prolate spheroidal spaces" are strictly band-limited to a given frequency band $[-F/2, F/2]$ and, among all spaces thus band-limited, maximally concentrated with respect to time in the sense that they assume a maximum part of their energy inside a given time interval $[-T/2, T/2]$ (see also Sections 3.4.2 and 4.6.2).

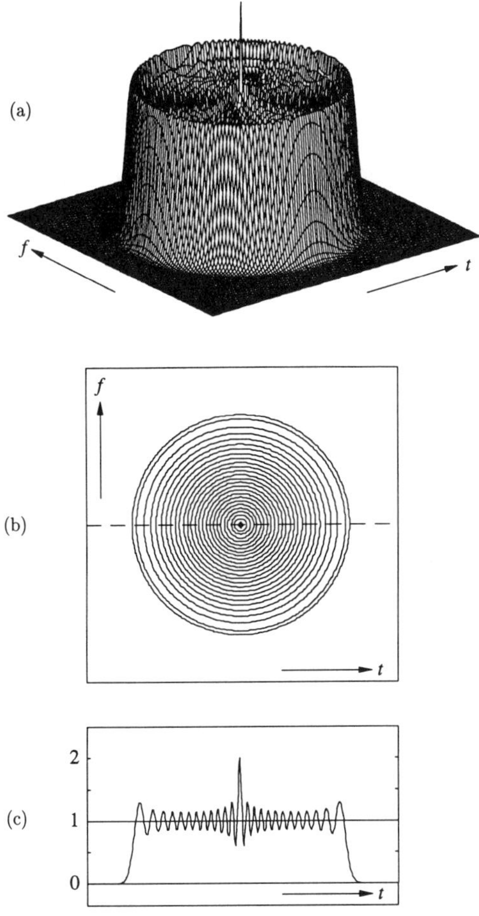

Figure 2.5. WD of Hermite space $\mathcal{H}_{25}^{(T)}$: (a) 3-D plot, (b) contour-line plot, and (c) cross section along the time axis $(f = 0)$.

We define the N-dimensional prolate spheroidal space

$$\mathcal{P}_N^{(T,F)} \triangleq \mathrm{span}\{p_k^{(T,F)}(t)\}_{k=1}^N$$

as the space spanned by the first N prolate spheroidal wave functions $p_k^{(T,F)}(t)$ [Slepian and Pollak, 1961, Landau and Pollak, 1961, Papoulis, 1984b]. These

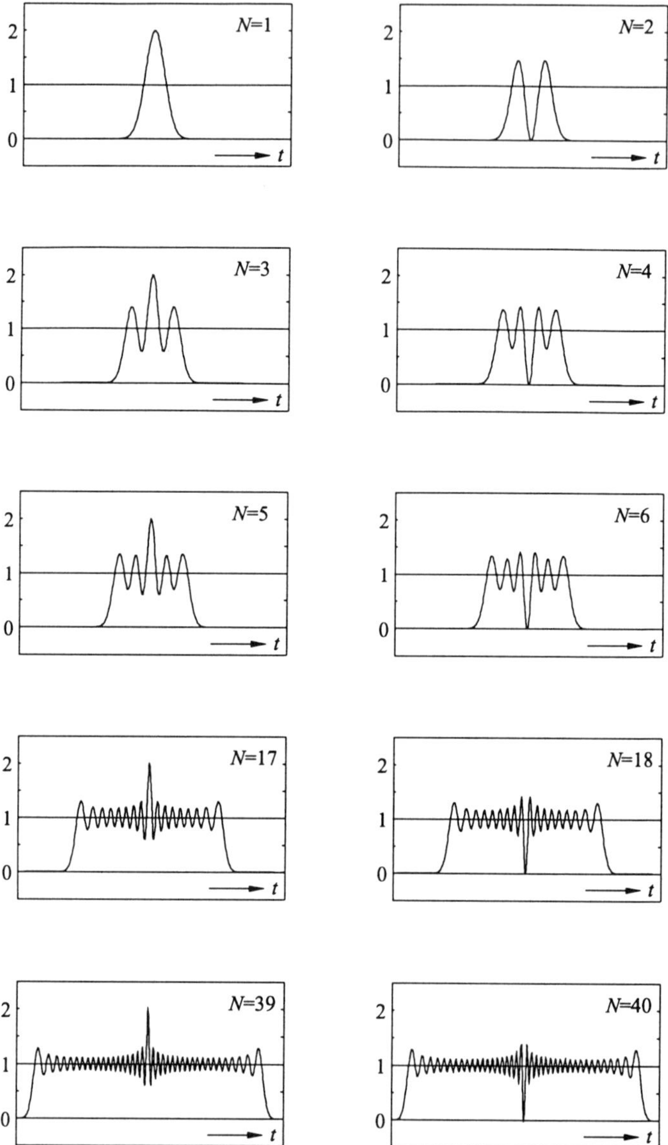

Figure 2.6. Cross sections of $W_{\mathcal{H}_N^{(T)}}(t, f)$ along the time axis ($f = 0$) for various dimensions N.

functions are orthonormal; they are the solutions to the eigenequation

$$\left(\mathbf{K}^{(T,F)}p_k^{(T,F)}\right)(t) \;=\; \lambda_k^{(T,F)}\,p_k^{(T,F)}(t)\,, \qquad k = 1, 2, \ldots \qquad (2.47)$$

corresponding to the N largest eigenvalues $\lambda_k^{(T,F)}$. Here, $\mathbf{K}^{(T,F)}$ is the composition of the time-limitation operator on $[-T/2, T/2]$ with the band-limitation operator (idealized lowpass filter) on $[-F/2, F/2]$, i.e.,

$$\left(\mathbf{K}^{(T,F)}\,x\right)(t) \;=\; F \int_{-T/2}^{T/2} \operatorname{sinc}\!\left(\pi F(t-t')\right)\,x(t')\,dt'$$

with $\operatorname{sinc}(\beta) = \frac{\sin(\beta)}{\beta}$, and the prolate spheroidal wave functions $p_k^{(T,F)}(t)$ are the normalized eigenfunctions of $\mathbf{K}^{(T,F)}$. Although $\mathbf{K}^{(T,F)}$ is not self-adjoint, the eigenvalues $\lambda_k^{(T,F)}$ are real-valued with $0 < \lambda_k^{(T,F)} < 1$, and the eigenfunctions $p_k^{(T,F)}(t)$ form an orthonormal basis of $\mathcal{F}[-F/2, F/2]$, the linear space of all signals band-limited in $[-F/2, F/2]$ [Slepian and Pollak, 1961, Landau and Pollak, 1961, Papoulis, 1984b]. Thus, any signal band-limited to $[-F/2, F/2]$ can be expanded into the prolate spheroidal wave functions $p_k^{(T,F)}(t)$. Since the $p_k^{(T,F)}(t)$ are band-limited to $[-F/2, F/2]$, the prolate spheroidal spaces $\mathcal{P}_N^{(T,F)}$ are subspaces of $\mathcal{F}[-F/2, F/2]$,

$$\mathcal{P}_N^{(T,F)} \;\subseteq\; \mathcal{F}[-F/2, F/2]\,.$$

From the finite-support property (2.15), it then follows that

$$W_{\mathcal{P}_N^{(T,F)}}(t, f) = 0 \quad \text{for} \quad |f| > F/2\,.$$

Fig. 2.7 shows the WD of a specific prolate spheroidal space $\mathcal{P}_N^{(T,F)}$ with $TF = 9.6$ and dimension $N = 10$. **Fig. 2.8** compares the WDs of several prolate spheroidal spaces $\mathcal{P}_N^{(T,F)}$ with fixed T and F and various dimensions N. Finally, **Fig. 2.9** compares the WDs of several prolate spheroidal spaces $\mathcal{P}_N^{(T,F)}$ with fixed ratio T/F, for various values of N and TF.

The prolate spheroidal spaces will be further considered in Sections 3.4.2, 4.6.2, 7.3, and 8.3.2.

2.5 Energetic Interpretation

From the way the WD of a space was defined in Section 2.1 by (loosely speaking) adding the WDs of all signals of the space, it is clear that the WD of a space can be interpreted as a "TF energy distribution" of the space under analysis. This interpretation is based directly on the energetic interpretation

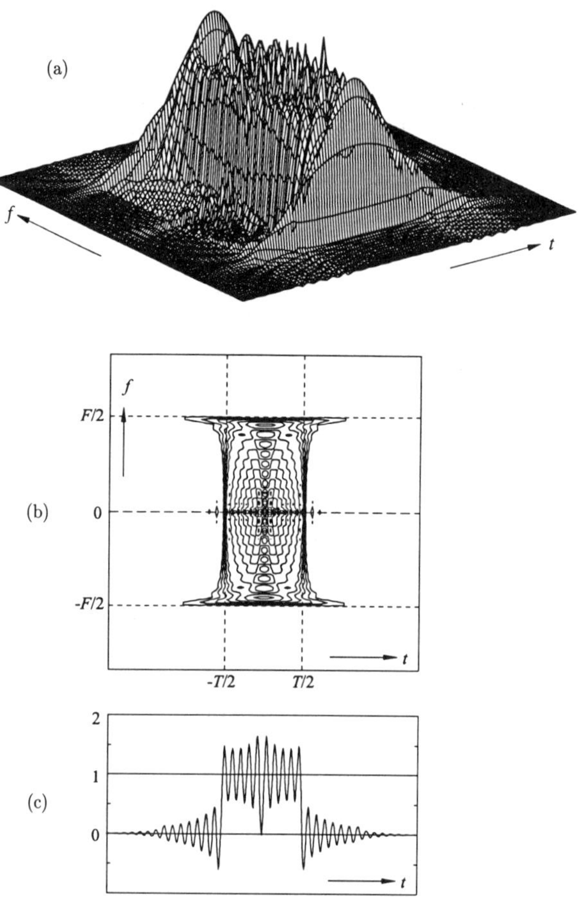

Figure 2.7.   WD of prolate spheroidal space $\mathcal{P}_N^{(T,F)}$ with $TF = 9.6$ and $N = 10$: (a) 3-D plot, (b) contour-line plot, and (c) cross section along the time axis ($f = 0$).

of the WD of a *signal*. On the other hand, it is well known that a *pointwise* energetic interpretation of the WD of a signal is not possible, since the uncertainty principle [Papoulis, 1984b, de Bruijn, 1967, Folland and Sitaram, 1997] prohibits the interpretation of any TF signal representation as a pointwise "TF energy density" [Claasen and Mecklenbräuker, 1984]. This restriction must then exist also for TF space representations such as the WD of a linear

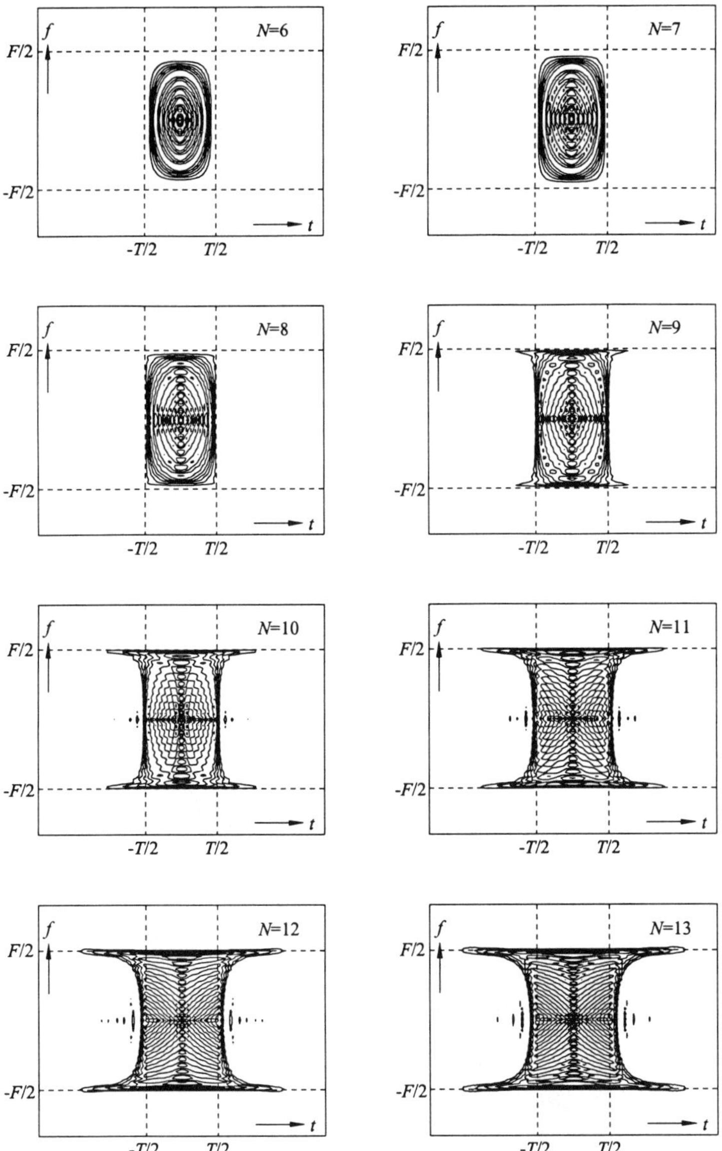

Figure 2.8. WD of prolate spheroidal spaces $\mathcal{P}_N^{(T,F)}$ with fixed T and F (with $TF = 9.6$), for various dimensions N.

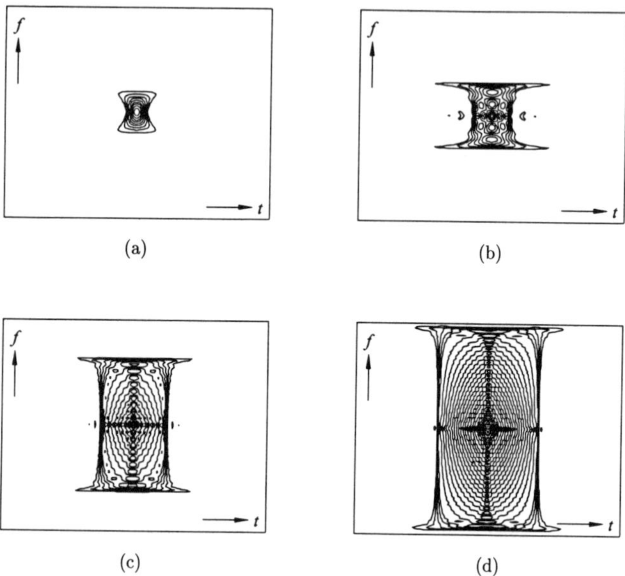

Figure 2.9.   WD of prolate spheroidal spaces $\mathcal{P}_N^{(T,F)}$ with fixed ratio $T/F = 100\,T_s^2$ (T_s is the sampling period), for various N and various TF: (a) $N = 1$, $TF = 1.44$, (b) $N = 3$, $TF = 2.56$, (c) $N = 10$, $TF = 10.24$, (d) $N = 23$, $TF = 23.04$.

signal space. However, it is still possible to maintain a loose interpretation of the WD of a space as an energy distribution, in the sense that certain local or global integrals of the WD of a space are related to well-defined energetic space quantities.

The basic energetic quantities of a space $\mathcal{X}$ are the energy/dimension $E_{\mathcal{X}} = \sum_{k=1}^{N_{\mathcal{X}}} E_{x_k} = \int_t P_{\mathcal{X}}(t,t)\,dt = N_{\mathcal{X}}$, the temporal energy density (instantaneous power) $d_{\mathcal{X}}(t) = \sum_{k=1}^{N_{\mathcal{X}}} d_{x_k}(t) = P_{\mathcal{X}}(t,t)$, and the spectral energy density $D_{\mathcal{X}}(f)$ $= \sum_{k=1}^{N_{\mathcal{X}}} D_{x_k}(f) = \tilde{P}_{\mathcal{X}}(f,f)$. Accordingly, the energetic interpretation of the WD of a space is primarily based on the *marginal properties* (2.25)–(2.27). These state that the energy densities of a signal space, $d_{\mathcal{X}}(t)$ and $D_{\mathcal{X}}(f)$, are "marginal densities" of the WD of the signal space, and that the space's total energy, $E_{\mathcal{X}} = N_{\mathcal{X}}$, is the WD's integral over the entire TF plane.

Apart from the marginal properties, there exists a less obvious energetic interpretation of the WD of a space that is phrased in terms of *local* WD integrals. This interpretation is related to a simple method for measuring the local energy content of a space $\mathcal{X}$ about a given TF point (t, f) [Hlawatsch and

Kozek, 1993]. Let $h(t')$ be a normalized signal which is well concentrated about the origin of the TF plane (e.g., a Gaussian signal). We form the "test signal"

$$h^{(t,f)}(t') \overset{\triangle}{=} h(t'-t)\, e^{j2\pi ft'}$$

by TF shifting $h(t')$ to the TF point (t, f). Clearly, $h^{(t,f)}(t')$ is well concentrated about (t, f). Hence, the energy of the projection of $h^{(t,f)}(t')$ onto $\mathcal{X}$, $\left\| h_{\mathcal{X}}^{(t,f)} \right\|^2$, will measure the amount of energy of $\mathcal{X}$ that is located in a local neighborhood of (t, f). Denoting this energy by $S_{\mathcal{X}}^{(h)}(t, f)$, we obtain

$$\begin{aligned} S_{\mathcal{X}}^{(h)}(t, f) \;\overset{\triangle}{=}\; & \left\| h_{\mathcal{X}}^{(t,f)} \right\|^2 \;=\; \langle W_{h^{(t,f)}}, W_{\mathcal{X}} \rangle \\ =\; & \int_{t'} \int_{f'} W_h(t'-t, f'-f)\, W_{\mathcal{X}}(t', f')\, dt'\, df', \end{aligned} \qquad (2.48)$$

where we have used the Moyal-type relation in (2.20) and the TF shift covariance property (2.33). Note that $0 \le S_{\mathcal{X}}^{(h)}(t, f) \le 1$ since $S_{\mathcal{X}}^{(h)}(t, f) = \left\| h_{\mathcal{X}}^{(t,f)} \right\|^2$ and $h(t')$ was assumed normalized.

We may view $S_{\mathcal{X}}^{(h)}(t, f)$ as a (nonnegative) TF representation of the space $\mathcal{X}$ which assigns to each TF point (t, f) the local energy content of $\mathcal{X}$ about (t, f). Due to (2.48), this local energy content is actually a local average[4] of the WD $W_{\mathcal{X}}(t, f)$ or, equivalently, $S_{\mathcal{X}}^{(h)}(t, f)$ is simply a smoothed version of $W_{\mathcal{X}}(t, f)$. Thus, *whereas a pointwise energetic interpretation of the WD of a space is impossible, local averages of the WD of a space do permit an energetic interpretation.* Also, while the WD of a space may be locally negative, suitably defined local integrals (or averages) are always nonnegative. All this is analogous to the WD of a signal [Claasen and Mecklenbräuker, 1980, Claasen and Mecklenbräuker, 1980b, Janssen, 1997b]. **Fig. 2.10** illustrates the local averaging (or smoothing) of the WD as expressed by (2.48). This smoothing is seen to suppress all local oscillations contained in the WD.

We note that the nonnegative TF representation $S_{\mathcal{X}}^{(h)}(t, f)$ is in fact the *spectrogram of the space* $\mathcal{X}$ as previously defined in Section 2.2.2 (see (2.11) and (2.29)). An analogous interpretation can also be given for the scalogram of a space (see (2.12) and (2.30)).

2.6 Extensions

This section considers two extensions of the WD of a signal space, namely, the cross-WD and a discrete-time WD version [Hlawatsch and Kozek, 1993].

---

[4]Note that the weighting function used for the local average (2.48), $W_h(t, f)$, is properly normalized due to the assumption that $h(t')$ is normalized: $\int_t \int_f W_h(t, f)\, dt\, df = \|h\|^2 = 1$.

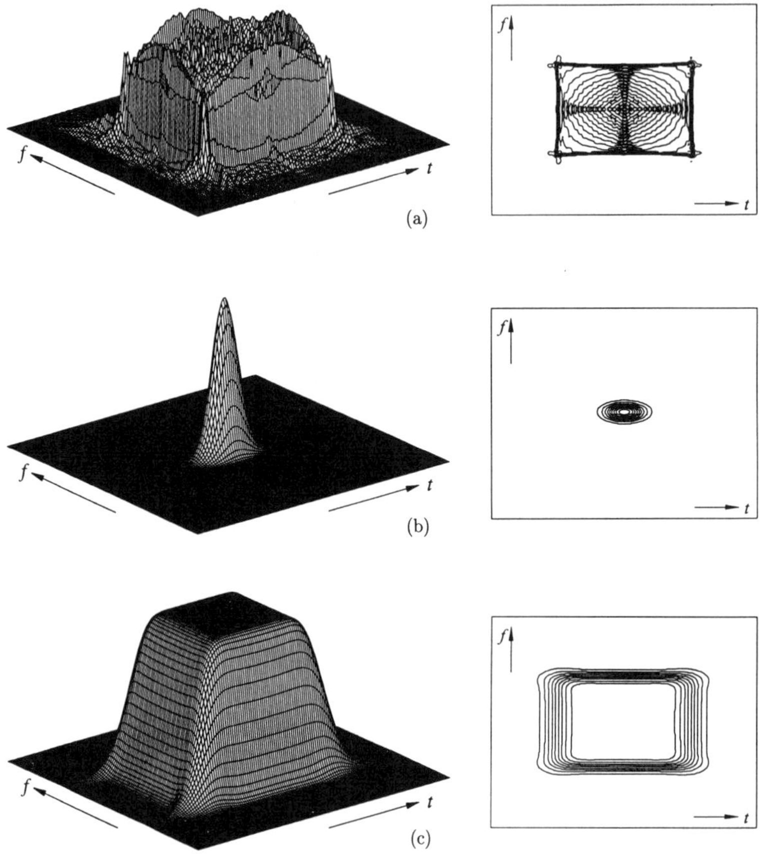

Figure 2.10. Spectrogram of a signal space—interpretation as smoothed WD: (a) WD of signal space $\mathcal{X}$, (b) WD of test signal (window) $h(t)$, (c) spectrogram of signal space $\mathcal{X}$.

2.6.1 Cross-Wigner Distribution

The cross-WD (CWD) $W_{\mathcal{X},\mathcal{Y}}(t,f)$ of two signal spaces $\mathcal{X}$ and $\mathcal{Y}$ can be defined by extending the equivalent definitions (2.1) and (2.3) of the auto-WD,

$$W_{\mathcal{X},\mathcal{Y}}(t,f) \;\triangleq\; \sum_{k=1}^{\infty} W_{l_{k,\mathcal{X}},l_{k,\mathcal{Y}}}(t,f) \;=\; \mathrm{E}\{W_{w_{\mathcal{X}},w_{\mathcal{Y}}}(t,f)\}\,.$$

Here,

$$W_{x,y}(t,f) \;=\; \int_{\tau} x\!\left(t+\frac{\tau}{2}\right) y^{*}\!\left(t-\frac{\tau}{2}\right) e^{-j2\pi f\tau}\, d\tau \tag{2.49}$$

is the CWD of two signals $x(t)$, $y(t)$ [Claasen and Mecklenbräuker, 1980]. Like the CWD of two signals, the CWD of two signal spaces is generally complex-valued, and it satisfies

$$W_{\mathcal{Y},\mathcal{X}}(t,f) = W_{\mathcal{X},\mathcal{Y}}^*(t,f) \qquad \text{and} \qquad W_{\mathcal{X},\mathcal{X}}(t,f) = W_{\mathcal{X}}(t,f) \,.$$

The CWD can be expressed in terms of the spaces' orthogonal projection operators $\mathbf{P}_{\mathcal{X}}$ and $\mathbf{P}_{\mathcal{Y}}$ or orthonormal bases $\{x_k(t)\}_{k=1}^{N_{\mathcal{X}}}$ and $\{y_l(t)\}_{l=1}^{N_{\mathcal{Y}}}$ as

$$
\begin{aligned}
W_{\mathcal{X},\mathcal{Y}}(t,f) &= \int_{\tau} (\mathbf{P}_{\mathcal{X}}\mathbf{P}_{\mathcal{Y}}^{+})\Big(t+\frac{\tau}{2},t-\frac{\tau}{2}\Big)\, e^{-j2\pi f\tau}\, d\tau \\
&= \sum_{k=1}^{N_{\mathcal{X}}} \sum_{l=1}^{N_{\mathcal{Y}}} \langle y_l, x_k\rangle\, W_{x_k,y_l}(t,f)
\end{aligned}
\tag{2.50}
$$

(cf. (2.4) and (2.7)). The last expression can also be written as

$$W_{\mathcal{X},\mathcal{Y}}(t,f) = \sum_{k=1}^{N_{\mathcal{X}}} W_{x_k,\, x_{k,\mathcal{Y}}}(t,f) = \sum_{l=1}^{N_{\mathcal{Y}}} W_{y_{l,\mathcal{X}},\, y_l}(t,f) \,,$$

where $x_{k,\mathcal{Y}}(t)$ and $y_{l,\mathcal{X}}(t)$ denote the orthogonal projections of $x_k(t)$ onto $\mathcal{Y}$ and of $y_l(t)$ onto $\mathcal{X}$, respectively.

Many of the properties formulated for the auto-WD in Section 2.3 can be extended to the CWD. Two further remarkable properties of the CWD are

$$W_{\mathcal{X},\mathcal{Y}}(t,f) \equiv 0 \quad \Longleftrightarrow \quad \mathcal{X} \perp \mathcal{Y} \tag{2.51}$$

and

$$W_{\mathcal{X},\mathcal{Y}}(t,f) = W_{\mathcal{X}}(t,f) \quad \Longleftrightarrow \quad \mathcal{X} \subseteq \mathcal{Y} \,. \tag{2.52}$$

In particular, we conclude from (2.51) that the CWD of two TF disjoint (and, hence, orthogonal—see Section 3.2.2) spaces is identically zero.

Fig. 2.11 shows the CWD of two TF shifted Hermite spaces. We observe the interesting fact that, in this example, the effective TF support of the CWD approximately equals the TF region where the corresponding auto-WDs overlap. This behavior is consistent with properties (2.51) and (2.52).

It should be noted that the CWD of two spaces is only partly analogous to the CWD of two signals. Indeed, while the CWD of two signals naturally occurs in the "quadratic superposition law"

$$W_{x+y}(t,f) = W_x(t,f) + W_y(t,f) + 2\,\text{Re}\{W_{x,y}(t,f)\} \,,$$

a similar expression does not generally exist for the WD $W_{\mathcal{X}+\mathcal{Y}}(t,f)$ of the sum of two spaces. Furthermore, as shown by (2.50), the CWD of two one-dimensional spaces $\mathcal{X}$, $\mathcal{Y}$ is generally *not* equal to the CWD of the (normalized)

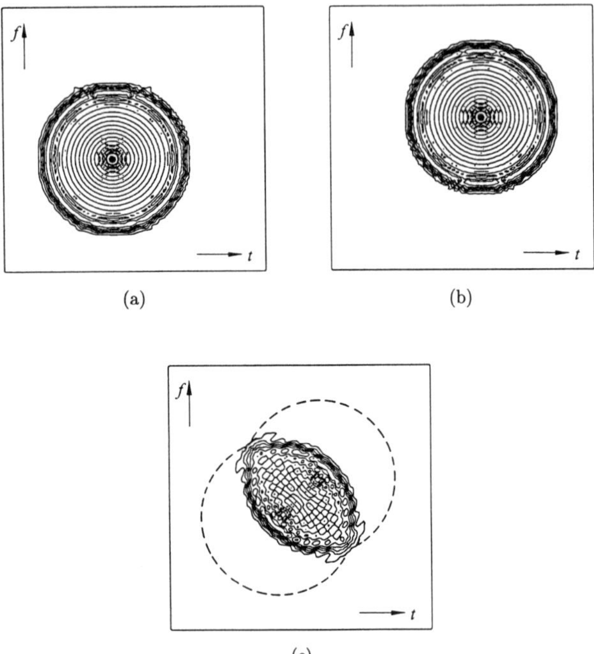

Figure 2.11.   CWD of two TF shifted Hermite spaces $\mathcal{X}$, $\mathcal{Y}$ (both with dimension 16): (a) auto-WD of space $\mathcal{X}$, (b) auto-WD of space $\mathcal{Y}$, (c) magnitude of CWD of spaces $\mathcal{X}$, $\mathcal{Y}$ with effective TF supports of the corresponding auto-WDs indicated by broken lines.

basis signals $x_1(t)$, $y_1(t)$. In particular, the CWD of two (nonzero) orthogonal *signals* is not identically zero whereas due to (2.51) the CWD of two orthogonal spaces is identically zero.

2.6.2   Discrete-Time Wigner Distribution

For a practical implementation of WD analysis on a digital computer, the WD of a signal space must be reformulated in a discrete-time setting. We therefore consider a linear space $\mathcal{X}$ of discrete-time signals $x(n)$ ($n = 0, \pm 1, \pm 2, ...$) which is characterized by its orthogonal projection operator $\mathbf{P}_{\mathcal{X}}$ with kernel $P_{\mathcal{X}}(n, n')$ or, alternatively, by an orthonormal[5] basis $\{x_k(n)\}_{k=1}^{N_{\mathcal{X}}}$.

---

[5]The inner product of two discrete-time signals $x(n)$ and $y(n)$ is defined as $\langle x, y \rangle = \sum_n x(n)\, y^*(n)$. A basis $\{x_k(n)\}$ is orthonormal if $\langle x_k, x_l \rangle = \delta_{kl}$.

Using the discrete-time WD of a signal $x(n)$ [Claasen and Mecklenbräuker, 1980a],

$$W_x^{(D)}(n, \theta) = 2 \sum_m x(n+m)\, x^*(n-m)\, e^{-j4\pi\theta m}, \qquad (2.53)$$

it is straightforward to reformulate the definitions (2.1) and (2.3) of the WD of a space in the discrete-time setting considered here. We then obtain the following expressions for the discrete-time WD of a discrete-time space $\mathcal{X}$:

$$W_{\mathcal{X}}^{(D)}(n, \theta) = 2 \sum_m P_{\mathcal{X}}(n+m,\, n-m)\, e^{-j4\pi\theta m} \qquad (2.54)$$

$$= \sum_{k=1}^{N_{\mathcal{X}}} W_{x_k}^{(D)}(n, \theta) \qquad (2.55)$$

(cf. (2.5), (2.7)). Like the discrete-time WD of a signal in (2.53), $W_{\mathcal{X}}^{(D)}(n, \theta)$ is periodic, with respect to the normalized-frequency variable θ, with period $1/2$, i.e., one half of the spectral period 1 of discrete-time signals. This will cause aliasing effects unless all $x(n) \in \mathcal{X}$ are band-limited to some fixed "halfband" $|\theta - \theta_0| < 1/4$ with bandwidth $1/2$ and given (application-dependent) center frequency θ_0. In other words, aliasing is avoided if $\mathcal{X} \subseteq \mathcal{H}^{(\theta_0)}$, where $\mathcal{H}^{(\theta_0)}$ denotes the "halfband subspace" of all signals satisfying the above band-limitation property with fixed θ_0. In particular, $\theta_0 = 0$ corresponds to a halfband containing (among other signals) all real-valued signals oversampled by a factor of 2 while $\theta_0 = 1/4$ corresponds to analytic signals.

Let us make these statements mathematically precise. To a linear signal space $\mathcal{X}_c$ of continuous-time signals $x_c(t)$, we can associate a linear signal space $\mathcal{X}$ of discrete-time signals $x(n)$ as

$$\mathcal{X} \triangleq \left\{ x(n) \;:\; x(n) = x_c(nT) \text{ with } x_c(t) \in \mathcal{X}_c \right\}$$

where $T > 0$ is a given sampling period. Thus, the space $\mathcal{X}$ is made up by the sampled versions of all elements $x_c(t)$ of $\mathcal{X}_c$. We assume that $\mathcal{X}_c$ is band-limited with total bandwidth F (this includes the negative frequency band in the case of a real-valued signal), and that $T \leq 1/F$ so that aliasing is avoided. Then, the following is easily verified: If $\{x_{c,k}(t)\}$ is an orthonormal basis of $\mathcal{X}_c$, then $\{x_k(n)\}$ with $x_k(n) = \sqrt{T}\, x_{c,k}(nT)$ will be an orthonormal basis of $\mathcal{X}$. Note that the dimensions of the spaces $\mathcal{X}_c$ and $\mathcal{X}$ are identical. Using (2.55) and (cf. [Claasen and Mecklenbräuker, 1980a])

$$W_{x_k}^{(D)}(n, \theta) = \sum_{l=-\infty}^{\infty} W_{x_{c,k}}\left(nT, \frac{\theta - l/2}{T}\right) \quad \text{for } x_k(n) = \sqrt{T}\, x_{c,k}(nT),$$

it follows that the discrete-time WD of $\mathcal{X}$ is a time-sampled and frequency-periodized version of the continuous-time WD of $\mathcal{X}_c$,

$$W_{\mathcal{X}}^{(D)}(n,\theta) = \sum_{l=-\infty}^{\infty} W_{\mathcal{X}_c}\left(nT, \frac{\theta - l/2}{T}\right). \tag{2.56}$$

The frequency periodization may cause aliasing in the WD. WD aliasing will however be avoided if the total bandwidth of $\mathcal{X}_c$ satisfies $F \leq f_s/2$, where $f_s = 1/T$ is the sampling rate. Here, the resulting discrete-time space $\mathcal{X}$ is a subspace of some halfband space $\mathcal{H}^{(\theta_0)}$ as discussed above, and (2.56) simplifies to

$$W_{\mathcal{X}}^{(D)}(n,\theta) = W_{\mathcal{X}_c}\left(nT, \frac{\theta}{T}\right) \qquad \text{for } \theta \in [\theta_0 - 1/4, \theta_0 + 1/4].$$

Hence, inside the halfband's fundamental frequency interval $|\theta - \theta_0| < 1/4$, the discrete-time WD is a sampled, *non-aliased* version of the continuous-time WD.

According to (2.55), $W_{\mathcal{X}}^{(D)}(n,\theta)$ can be calculated simply by adding the WDs of all basis signals $x_k(n)$. A more efficient implementation is obtained by first forming the projection operator kernel $P_{\mathcal{X}}(n,n') = \sum_{k=1}^{N_x} x_k(n)\, x_k^*(n')$ and then applying the FFT to compute a discrete-frequency version of the Fourier transform (2.54).

2.7 Minimization/Maximization of Time-Frequency Integrals

In subsequent chapters, we shall repeatedly encounter the problem of finding the linear signal space $\mathcal{X}$ minimizing or maximizing a weighted integral of the WD [de Bruijn, 1967, Janssen, 1989, Janssen, 1997b],

$$Q_{\mathcal{X}} = \int_t \int_f W_{\mathcal{X}}(t,f)\, G(t,f)\, dt\, df = \langle W_{\mathcal{X}}, G \rangle, \tag{2.57}$$

with real-valued TF weighting function $G(t,f)$. This problem is directly relevant to the uncertainty relations for signal spaces presented in Sections 3.3 and 3.4, the TF synthesis of signal spaces discussed in Chapter 4, and the optimum projection filters for signal estimation and detection derived in Chapter 6.

We shall first establish an alternative expression for $Q_{\mathcal{X}}$ by representing $G(t,f)$ as the Weyl symbol [Kozek, 1992b, Folland, 1989, Janssen, 1989, Shenoy and Parks, 1994, Ramanathan and Topiwala, 1993, Kozek and Hlawatsch, 1991b, Kozek, 1992a] of a linear operator $\mathbf{H}$ with kernel $H(t_1, t_2)$:

$$G(t,f) = \int_\tau H\left(t + \frac{\tau}{2}, t - \frac{\tau}{2}\right) e^{-j2\pi f\tau}\, d\tau, \tag{2.58}$$

$$H(t_1, t_2) = \int_f G\left(\frac{t_1 + t_2}{2}, f\right) e^{j2\pi(t_1 - t_2)f}\, df. \tag{2.59}$$

Then, $Q_{\mathcal{X}}$ can be written as

$$Q_{\mathcal{X}} = \int_{t_1} \int_{t_2} H(t_1, t_2)\, P_{\mathcal{X}}^*(t_1, t_2)\, dt_1 dt_2 = \sum_{k=1}^{N_{\mathcal{X}}} Q_{x_k} \tag{2.60}$$

where $\{x_k(t)\}_{k=1}^{N_{\mathcal{X}}}$ is an orthonormal basis of $\mathcal{X}$ and

$$Q_{x_k} = \langle W_{x_k}, G \rangle = \int_{t_1} \int_{t_2} H(t_1, t_2)\, x_k^*(t_1)\, x_k(t_2)\, dt_1 dt_2 = \langle \mathbf{H} x_k, x_k \rangle$$

is recognized as the quadratic form of $x_k(t)$ generated by the operator $\mathbf{H}$. Hence, $Q_{\mathcal{X}}$ can also be written as the sum of quadratic forms

$$Q_{\mathcal{X}} = \sum_{k=1}^{N_{\mathcal{X}}} \langle \mathbf{H} x_k, x_k \rangle . \tag{2.61}$$

Since $G(t, f)$ is real-valued, $\mathbf{H}$ is a self-adjoint operator, i.e., its kernel $H(t_1, t_2)$ is Hermitian in the sense that $H^*(t_2, t_1) = H(t_1, t_2)$.

In what follows, we shall assume that the operator $\mathbf{H}$ has a discrete spectrum[6], i.e., a discrete set of eigenvalues λ_k and eigenfunctions $u_k(t)$ defined by the eigenequation

$$\big(\mathbf{H} u_k \big)(t) = \lambda_k\, u_k(t) .$$

Since $\mathbf{H}$ is a self-adjoint operator, the eigenvalues λ_k are real-valued and the eigenfunctions $u_k(t)$ (suitably augmented if necessary) form an orthonormal basis of $\mathcal{L}_2(\mathbb{R})$. It is easily shown that, for $x(t) = u_k(t)$, the WD integral Q_x becomes the eigenvalue λ_k,

$$Q_{u_k} = \langle W_{u_k}, G \rangle = \langle \mathbf{H} u_k, u_k \rangle = \lambda_k . \tag{2.62}$$

We note that this can be generalized as

$$\langle W_{u_k, u_l}, G \rangle = \langle \mathbf{H} u_k, u_l \rangle = \lambda_k\, \delta_{kl} \tag{2.63}$$

where $W_{u_k, u_l}(t, f)$ is the cross-WD of $u_k(t)$ and $u_l(t)$ (see (2.49)) and δ_{kl} is the Kronecker delta symbol. Inserting the spectral representation of $\mathbf{H}$,

$$H(t_1, t_2) = \sum_{k=1}^{\infty} \lambda_k\, u_k(t_1)\, u_k^*(t_2) ,$$

---

[6] A sufficient condition for the existence of a discrete spectrum is the square-integrability of $G(t, f)$, which entails the square-integrability of $H(t_1, t_2)$ and, in turn, the compactness of the operator $\mathbf{H}$ [Naylor and Sell, 1982].

into (2.58), we obtain the relation [Hlawatsch and Krattenthaler, 1989, Janssen, 1989, Hlawatsch et al., 1990]

$$G(t, f) = \sum_{k=1}^{\infty} \lambda_k \, W_{u_k}(t, f) \tag{2.64}$$

which expresses the weighting function $G(t, f)$ as a linear combination of the WDs $W_{u_k}(t, f)$ of the orthonormal eigenfunctions $u_k(t)$.

2.7.1 Unconstrained Minimization/Maximization

The next theorem, whose proof is provided in Appendix 2.A, characterizes the linear signal space that minimizes or maximizes the weighted WD integral $Q_{\mathcal{X}}$ subject to prescribed space dimension $N_{\mathcal{X}} = N$. Apart from the given dimension, the space is not constrained any further. In Section 2.7.2, we shall solve the minimization/maximization problem under a subspace constraint.

Theorem 2.1 *Let $G(t, f)$ be a real-valued weighting function whose corresponding operator* **H** *has a discrete spectrum of eigenvalues λ_k and (orthonormal) eigenfunctions $u_k(t)$, and let $Q_{\mathcal{X}} = \langle W_{\mathcal{X}}, G \rangle$ be the weighted WD integral induced by $G(t, f)$.*

Case 1. If the eigenvalues λ_k can be arranged in non-decreasing order as $\lambda_1 \leq \lambda_2 \leq ...$, i.e., if there exists a minimum eigenvalue, then there exists an N-dimensional space minimizing the WD integral $Q_{\mathcal{X}}$,

$$\mathcal{X}_{\mathrm{opt}}(N) \;\overset{\triangle}{=}\; \arg \min_{N_{\mathcal{X}}=N} \, Q_{\mathcal{X}} \,.$$

This space is spanned by the first N eigenfunctions $u_k(t)$ of **H**, *i.e., the eigenfunctions corresponding to the N smallest eigenvalues λ_k,*

$$\mathcal{X}_{\mathrm{opt}}(N) \;=\; \mathrm{span}\{u_k(t)\}_{k=1}^{N} \,.$$

The minimum WD integral (minimum for the prescribed space dimension N) is given by the sum of the first N (i.e., the N smallest) eigenvalues,

$$Q_{\mathcal{X},\mathrm{min}}\Big|_{N_{\mathcal{X}}=N} \;=\; Q_{\mathcal{X}_{\mathrm{opt}}(N)} \;=\; \sum_{k=1}^{N} \lambda_k \,.$$

Case 2. If the eigenvalues λ_k can be arranged in non-increasing order as $\lambda_1 \geq \lambda_2 \geq ...$, i.e., if there exists a maximum eigenvalue, then there exists an N-dimensional space maximizing the WD integral $Q_{\mathcal{X}}$,

$$\mathcal{X}_{\mathrm{opt}}(N) \;\overset{\triangle}{=}\; \arg \max_{N_{\mathcal{X}}=N} \, Q_{\mathcal{X}} \,.$$

This space is spanned by the first N eigenfunctions $u_k(t)$ of $\mathbf{H}$, i.e., the eigenfunctions corresponding to the N largest eigenvalues λ_k,

$$\mathcal{X}_{\mathrm{opt}}(N) \;=\; \mathrm{span}\{u_k(t)\}_{k=1}^{N}\,.$$

The maximum WD integral (maximum for the prescribed space dimension N) is given by the sum of the first N (i.e., the N largest) eigenvalues,

$$Q_{\mathcal{X},\mathrm{max}}\Big|_{N_{\mathcal{X}}=N} \;=\; Q_{\mathcal{X}_{\mathrm{opt}}(N)} \;=\; \sum_{k=1}^{N}\lambda_k\,.$$

If 0 is an eigenvalue of $\mathbf{H}$ with multiplicity ≥ 1, and if 0 falls into the set of the N smallest or largest eigenvalues, then (an appropriate number of) the corresponding eigenfunctions must be included in the set of N eigenfunctions spanning the optimum space $\mathcal{X}_{\mathrm{opt}}(N)$.

In the above theorem, we have assumed that the dimension of the space is fixed beforehand. However, since the theorem gives the minimum or maximum value of $Q_{\mathcal{X}}$ for any given dimension N as the sum of the N smallest or largest eigenvalues, respectively, it is now easy to derive the optimum dimension N. In Case 1, it is readily seen that the absolutely minimum $Q_{\mathcal{X}}$ is achieved by the space spanned by all eigenfunctions $u_k(t)$ corresponding to *negative* eigenvalues. Hence, if $N_- > 0$ denotes the number of negative eigenvalues, then the absolutely optimum space is

$$\mathcal{X}_{\mathrm{opt}} \;\overset{\triangle}{=}\; \arg\min_{\mathcal{X}} Q_{\mathcal{X}} \;=\; \mathrm{span}\{u_k(t)\}_{k=1}^{N_-}\,,$$

and the absolutely minimum $Q_{\mathcal{X}}$ is

$$Q_{\mathcal{X},\mathrm{min}} \;=\; Q_{\mathcal{X}_{\mathrm{opt}}} \;=\; \sum_{k=1}^{N_-}\lambda_k \;<\; 0\,.$$

However, if there are no negative eigenvalues, i.e., if $N_- = 0$ (the operator $\mathbf{H}$ is positive semidefinite), then $\mathcal{X}_{\mathrm{opt}}$ is spanned by any set of eigenfunctions corresponding to zero eigenvalues and $Q_{\mathcal{X},\mathrm{min}} = 0$. If there are only positive eigenvalues ($\mathbf{H}$ is positive definite), then $\mathcal{X}_{\mathrm{opt}} = \mathcal{Z}$ (the "zero space" considered in Section 2.4) and $Q_{\mathcal{X},\mathrm{min}} = Q_{\mathcal{Z}} = 0$.

An analogous discussion applies to Case 2: here, the absolutely maximum $Q_{\mathcal{X}}$ is achieved by the space spanned by all eigenfunctions $u_k(t)$ corresponding to *positive* eigenvalues. Hence, if $N_+ > 0$ denotes the number of positive eigenvalues, then the absolutely optimum space is

$$\mathcal{X}_{\mathrm{opt}} \;\overset{\triangle}{=}\; \arg\max_{\mathcal{X}} Q_{\mathcal{X}} \;=\; \mathrm{span}\{u_k(t)\}_{k=1}^{N_+}\,,$$

and the absolutely maximum $Q_\mathcal{X}$ is

$$Q_{\mathcal{X},\max} = Q_{\mathcal{X}_{\mathrm{opt}}} = \sum_{k=1}^{N_+} \lambda_k > 0.$$

If there are no positive eigenvalues, i.e., if $N_+ = 0$ (**H** is negative semidefinite), then $\mathcal{X}_{\mathrm{opt}}$ is spanned by any set of eigenfunctions corresponding to zero eigenvalues and $Q_{\mathcal{X},\max} = 0$. If there are only positive eigenvalues (**H** is negative definite), then $\mathcal{X}_{\mathrm{opt}} = \mathcal{Z}$ and $Q_{\mathcal{X},\max} = Q_\mathcal{Z} = 0$.

2.7.2 Subspace-Constrained Minimization/Maximization

We shall now consider the minimization/maximization of the weighted WD integral $Q_\mathcal{X}$ under a *subspace constraint* $\mathcal{X} \subseteq \mathcal{C}$, i.e., the space $\mathcal{X}$ is constrained to be a subspace of a given signal space $\mathcal{C} \subseteq \mathcal{L}_2(\mathbb{R})$. The advantage of the subspace constraint is that it allows us to enforce certain properties of the space $\mathcal{X}$ (e.g. band-limitation in a given frequency band, see Sections 3.4.2, 4.3, and 4.6.2).

In the following, the *$\mathcal{C}$-restricted operator*

$$\mathbf{H}_\mathcal{C} \triangleq \mathbf{P}_\mathcal{C}\mathbf{H}\mathbf{P}_\mathcal{C},$$

where $\mathbf{P}_\mathcal{C}$ is the orthogonal projection operator on the constraint space $\mathcal{C}$, will play an important role. Like **H**, $\mathbf{H}_\mathcal{C}$ is a self-adjoint operator. For a space $\mathcal{X} \subseteq \mathcal{C}$, the orthonormal basis signals spanning $\mathcal{X}$ are elements of the constraint space $\mathcal{C}$, i.e., $x_k(t) \in \mathcal{C}$. Hence, the basis signals satisfy

$$\big(\mathbf{P}_\mathcal{C}\, x_k\big)(t) = x_k(t),$$

which implies that the WD integral $Q_\mathcal{X}$ can be written as

$$Q_\mathcal{X} = \sum_{k=1}^{N_\mathcal{X}} \langle \mathbf{H}x_k, x_k \rangle = \sum_{k=1}^{N_\mathcal{X}} \langle \mathbf{H}\mathbf{P}_\mathcal{C}\, x_k, \mathbf{P}_\mathcal{C}\, x_k \rangle = \sum_{k=1}^{N_\mathcal{X}} \langle \mathbf{H}_\mathcal{C}\, x_k, x_k \rangle. \qquad (2.65)$$

In what follows, we shall assume that the operator $\mathbf{H}_\mathcal{C}$ has a discrete spectrum of eigenvalues μ_k and eigenfunctions $v_k(t)$. Since $\mathbf{H}_\mathcal{C}$ is a self-adjoint operator, the eigenvalues μ_k are real-valued and the eigenfunctions $v_k(t)$ (suitably augmented if necessary) form an orthonormal basis of $\mathcal{C}$. Furthermore, we note that

$$Q_{v_k} = \langle \mathbf{H}_\mathcal{C}\, v_k, v_k \rangle = \mu_k, \qquad (2.66)$$

which can be generalized as

$$\langle \mathbf{H}_\mathcal{C}\, v_k, v_l \rangle = \mu_k\, \delta_{kl}.$$

The next theorem, proved in Appendix 2.B, characterizes the linear signal space $\mathcal{X} \subseteq \mathcal{C}$ that minimizes or maximizes the weighted WD integral $Q_\mathcal{X}$ subject to prescribed dimension $N_\mathcal{X} = N$. Since $\mathcal{X} \subseteq \mathcal{C}$, the dimension N prescribed may not be larger than the dimension $N_\mathcal{C}$ of the constraint space $\mathcal{C}$.

Theorem 2.2 *Let $G(t, f)$ be a real-valued weighting function corresponding to the operator* $\mathbf{H}$, *and let* $Q_\mathcal{X} = \langle W_\mathcal{X}, G \rangle$ *be the weighted WD integral induced by* $G(t, f)$. *Furthermore let* $\mathcal{C} \subseteq \mathcal{L}_2(\mathbb{R})$ *be a given linear signal space, and assume that the $\mathcal{C}$-restricted operator* $\mathbf{H}_\mathcal{C} = \mathbf{P}_\mathcal{C} \mathbf{H} \mathbf{P}_\mathcal{C}$ *has a discrete spectrum of eigenvalues* μ_k *and eigenfunctions* $v_k(t)$.

Case 1. If the eigenvalues μ_k can be arranged in non-decreasing order as $\mu_1 \leq \mu_2 \leq ...$, i.e., if there exists a minimum eigenvalue, then there exists an N-dimensional space (with $N \leq N_\mathcal{C}$) minimizing the WD integral $Q_\mathcal{X}$ under the subspace constraint $\mathcal{X} \subseteq \mathcal{C}$,

$$\mathcal{X}_{\mathrm{opt}}(N, \mathcal{C}) \triangleq \arg \min_{\substack{N_\mathcal{X} = N \\ \mathcal{X} \subseteq \mathcal{C}}} Q_\mathcal{X} \, .$$

This space is spanned by the first N eigenfunctions $v_k(t)$ of $\mathbf{H}_\mathcal{C}$, i.e., the eigenfunctions corresponding to the N smallest eigenvalues μ_k,

$$\mathcal{X}_{\mathrm{opt}}(N, \mathcal{C}) = \mathrm{span}\{v_k(t)\}_{k=1}^{N} \subseteq \mathcal{C} \, .$$

The minimum WD integral (minimum for the prescribed space dimension N) is given by the sum of the first N (i.e., the N smallest) eigenvalues,

$$Q_{\mathcal{X},\mathrm{min}} \bigg|_{\substack{N_\mathcal{X} = N \\ \mathcal{X} \subseteq \mathcal{C}}} = Q_{\mathcal{X}_{\mathrm{opt}}(N, \mathcal{C})} = \sum_{k=1}^{N} \mu_k \, .$$

Case 2. If the eigenvalues μ_k can be arranged in non-increasing order as $\mu_1 \geq \mu_2 \geq ...$, i.e., if there exists a maximum eigenvalue, then there exists an N-dimensional space (with $N \leq N_\mathcal{C}$) maximizing the WD integral $Q_\mathcal{X}$ under the subspace constraint $\mathcal{X} \subseteq \mathcal{C}$,

$$\mathcal{X}_{\mathrm{opt}}(N, \mathcal{C}) \triangleq \arg \max_{\substack{N_\mathcal{X} = N \\ \mathcal{X} \subseteq \mathcal{C}}} Q_\mathcal{X} \, .$$

This space is spanned by the first N eigenfunctions $v_k(t)$ of $\mathbf{H}_\mathcal{C}$, i.e., the eigenfunctions corresponding to the N largest eigenvalues μ_k,

$$\mathcal{X}_{\mathrm{opt}}(N, \mathcal{C}) = \mathrm{span}\{v_k(t)\}_{k=1}^{N} \subseteq \mathcal{C} \, .$$

The maximum WD integral (maximum for the prescribed space dimension N) is given by the sum of the first N (i.e., the N largest) eigenvalues,

$$Q_{\mathcal{X},\max}\Big|_{\substack{N_{\mathcal{X}}=N \\ \mathcal{X}\subseteq\mathcal{C}}} = Q_{\mathcal{X}_{\mathrm{opt}}(N,\mathcal{C})} = \sum_{k=1}^{N} \mu_k \, .$$

In the case where the dimension of the constraint subspace $\mathcal{C}$ is small, an alternative method for finding the optimum space $\mathcal{X}_{\mathrm{opt}}(N,\mathcal{C}) \subseteq \mathcal{C}$ is particularly efficient. This method assumes the availability of an (arbitrary) orthonormal basis of the constraint subspace $\mathcal{C}$. It is formulated in the next theorem whose proof can be found in Appendix 2.C.

Theorem 2.3 *Let $\{c_j(t)\}_{j=1}^{N_C}$ be an orthonormal basis of the constraint subspace $\mathcal{C}$. An orthonormal basis of the N-dimensional space $\mathcal{X}_{\mathrm{opt}}(N,\mathcal{C}) \subseteq \mathcal{C}$ minimizing or maximizing $Q_{\mathcal{X}}$ (depending on whether Case 1 or Case 2 is valid) is given by*

$$v_k(t) = \sum_{j=1}^{N_C} \alpha_{kj}\, c_j(t)\,, \qquad k = 1, ..., N\,,$$

where the coefficients α_{kj} are derived by the following steps:

1. *Form the Hermitian $N_C \times N_C$ matrix $\mathbf{\Gamma}$ with elements*

$$(\mathbf{\Gamma})_{ij} = \langle G, W_{c_i,c_j} \rangle = \langle \mathbf{H}c_j, c_i \rangle\,, \qquad i,j = 1, ..., N_C\,.$$

2. *Calculate the eigenvalues μ_k and the normalized eigenvectors $\mathbf{v}_k$ of $\mathbf{\Gamma}$. (We assume that the eigenvalues μ_k are arranged in non-decreasing order in Case 1 and in non-increasing order in Case 2.)*

3. *The optimum coefficients α_{kj} are given as*

$$\alpha_{kj} = (\mathbf{v}_k)_j\,, \qquad k = 1, ..., N\,,$$

where $(\mathbf{v}_k)_j$ denotes the jth element of the kth eigenvector $\mathbf{v}_k$.

We note that the μ_k and $v_k(t)$ in Theorem 2.3 are in fact the eigenvalues and eigenfunctions, respectively, of the $\mathcal{C}$-restricted operator $\mathbf{H}_{\mathcal{C}}$. Thus, the eigenvalues μ_k of $\mathbf{H}_{\mathcal{C}}$ equal the eigenvalues of the matrix $\mathbf{\Gamma}$, and the eigenfunctions $v_k(t)$ of $\mathbf{H}_{\mathcal{C}}$ are related to the eigenvectors $\mathbf{v}_k$ of $\mathbf{\Gamma}$ according to

$$v_k(t) = \sum_{j=1}^{N_C} (\mathbf{v}_k)_j\, c_j(t)\,, \qquad k = 1, ..., N\,, \quad j = 1, ..., N_C\,.$$

So far, we have assumed that the dimension of the space $\mathcal{X}_{\mathrm{opt}}(N,\mathcal{C})$ is fixed beforehand. However, since Theorem 2.2 yields the minimum or maximum value of $Q_\mathcal{X}$ for any given dimension N as the sum of the N smallest or largest eigenvalues, respectively, it is again easy to derive the optimum dimension N. The resulting "absolutely optimum" spaces are obtained in complete analogy to the unconstrained case discussed in Section 2.7.1.

Appendix 2.A: Proof of Theorem 2.1

We first consider Case 1 defined in Theorem 2.1, i.e., the eigenvalues λ_k of $\mathbf{H}$ can be arranged in non-decreasing order as $\lambda_1 \leq \lambda_2 \leq ...$, and the weighted WD integral $Q_\mathcal{X}$ is to be minimized subject to given space dimension N.

It suffices to determine an orthonormal basis $\{x_k(t)\}_{k=1}^N$ of the optimum space $\mathcal{X}$. With (2.61), the quantity to be minimized is

$$Q_\mathcal{X} = \sum_{k=1}^{N} \langle \mathbf{H}x_k, x_k \rangle . \tag{2.A.1}$$

The eigenfunctions $u_k(t)$ of $\mathbf{H}$, suitably augmented if necessary, form an orthonormal basis of $\mathcal{L}_2(\mathbb{R})$. Hence, the basis signals can be expanded as

$$x_k(t) = \sum_{l=1}^{\infty} a_{kl}\, u_l(t), \qquad k = 1,...,N . \tag{2.A.2}$$

Let us define the coefficient vectors $\mathbf{a}_k$ whose lth elements are a_{kl}. It is easily shown that orthonormality of the $x_k(t)$ implies orthonormality of the $\mathbf{a}_k$, i.e., $\mathbf{a}_k^H \mathbf{a}_l = \delta_{kl}$ for $k, l = 1, 2, ..., N$. Inserting (2.A.2) into (2.A.1), we obtain

$$Q_\mathcal{X} = \sum_{k=1}^{N} \sum_{l=1}^{\infty} \sum_{m=1}^{\infty} a_{kl}\, a_{km}^* \langle \mathbf{H}u_l, u_m \rangle = \sum_{k=1}^{N} \sum_{l=1}^{\infty} \sum_{m=1}^{\infty} a_{kl}\, a_{km}^* \lambda_l\, \delta_{lm}$$

$$= \sum_{k=1}^{N} \sum_{l=1}^{\infty} \lambda_l\, |a_{kl}|^2 ,$$

where (2.63) has been used. This can be written as the sum of quadratic forms

$$Q_\mathcal{X} = \sum_{k=1}^{N} \mathbf{a}_k^H \mathbf{\Lambda} \mathbf{a}_k , \tag{2.A.3}$$

where $\mathbf{\Lambda}$ is the diagonal eigenvalue matrix containing the eigenvalues λ_k in its diagonal. We have to minimize (2.A.3) under the orthonormality constraint

$\mathbf{a}_k^H \mathbf{a}_l = \delta_{kl}$. It suffices, however, to use a *normalization constraint* $\mathbf{a}_k^H \mathbf{a}_k = 1$ since, as will be seen presently, this automatically yields orthonormal vectors $\mathbf{a}_k$. Using Lagrange multipliers ν_k, the problem then amounts to the unconstrained maximization of

$$\sigma_a \;=\; \sum_{k=1}^{N} \mathbf{a}_k^H \mathbf{\Lambda} \mathbf{a}_k \;+\; \sum_{k=1}^{N} \nu_k \left(1 - \mathbf{a}_k^H \mathbf{a}_k\right) \;=\; \sum_{k=1}^{N} \mathbf{a}_k^H \left(\mathbf{\Lambda} - \nu_k \mathbf{I}\right) \mathbf{a}_k \;+\; \sum_{k=1}^{N} \nu_k \;.$$

Setting the gradient of σ_a with respect to $\mathbf{a}_i$ equal to zero yields the system of equations

$$\mathbf{\Lambda} \mathbf{a}_i \;=\; \nu_i \mathbf{a}_i \,, \quad i = 1, ..., N \;.$$

Hence, the coefficient vectors $\mathbf{a}_i$ must be normalized eigenvectors of $\mathbf{\Lambda}$. Assuming all eigenvalues λ_i to be distinct[7], the eigenvectors of the diagonal matrix $\mathbf{\Lambda}$ are the unit vectors $\mathbf{e}_i$. Note that, as claimed above, the orthonormality constraint is indeed satisfied. Let us adopt any specific choice of N eigenvectors, $\mathbf{a}_k = \mathbf{e}_{i_k}$. Inserting this into (2.A.2), we see that the basis signals are the N eigenfunctions $u_{i_k}(t)$,

$$x_k(t) \;=\; \sum_{l=1}^{\infty} (\mathbf{a}_k)_l \, u_l(t) \;=\; \sum_{l=1}^{\infty} (\mathbf{e}_{i_k})_l \, u_l(t) \;=\; \sum_{l=1}^{\infty} \delta_{i_k,l} \, u_l(t) \;=\; u_{i_k}(t) \,,$$

so that

$$Q_\mathcal{X} \;=\; \sum_{k=1}^{N} Q_{u_{i_k}} \;=\; \sum_{k=1}^{N} \lambda_{i_k} \,,$$

where (2.60) and (2.62) have been used. This is finally minimized by choosing the indices $i_k = k$, so that $Q_\mathcal{X} = \sum_{k=1}^{N} \lambda_{i_k}$ becomes the sum of the N *smallest* eigenvalues. Hence, we obtain

$$x_k(t) \;=\; u_k(t) \,, \quad k = 1, ..., N \qquad \text{and} \qquad Q_\mathcal{X} \;=\; \sum_{k=1}^{N} \lambda_k \,,$$

which proves the first part of Theorem 2.1. The second part (pertaining to Case 2, i.e., maximization of $Q_\mathcal{X}$) is proved in a fully analogous manner.

Appendix 2.B: Proof of Theorem 2.2

This proof is largely analogous to the proof of Theorem 2.1 given in Appendix 2.A. We first consider Case 1, i.e., the eigenvalues μ_k of $\mathbf{H}_C$ can be arranged in

---

[7]The assumption of distinct eigenvalues is easily shown to be unnecessary. Indeed, if some eigenvalues are equal, the eigenvectors may be rotated versions of the unit vectors $\mathbf{e}_i$. Although this yields different basis signals, the space spanned by this rotated basis is identical to the space spanned by the unit vectors.

non-decreasing order as $\mu_1 \leq \mu_2 \leq ...$, and $Q_{\mathcal{X}}$ is to be minimized subject to the constraints $\mathcal{X} \subseteq \mathcal{C}$ and $N_{\mathcal{X}} = N$.

In order to determine an orthonormal basis $\{x_k(t)\}_{k=1}^{N}$ of the optimum space $\mathcal{X} \subseteq \mathcal{C}$, we recall that $Q_{\mathcal{X}}$ can be expressed in terms of the basis functions as

$$Q_{\mathcal{X}} = \sum_{k=1}^{N} \langle \mathbf{H}_{\mathcal{C}} \, x_k, x_k \rangle$$

(see (2.65)). The eigenfunctions $v_k(t)$ of $\mathbf{H}_{\mathcal{C}}$, suitably augmented if necessary, form an orthonormal basis of $\mathcal{C}$. Hence, the basis signals can be expanded as

$$x_k(t) = \sum_{l=1}^{N_{\mathcal{C}}} a_{kl} \, v_l(t), \qquad k = 1, ..., N .$$

Let us define the coefficient vectors $\mathbf{a}_k$ whose lth elements are a_{kl}. Proceeding as in Appendix 2.A, we can show that the optimum coefficient vectors $\mathbf{a}_i$ must be normalized eigenvectors of the diagonal matrix $\mathbf{M}$ containing the eigenvalues μ_k in its diagonal,

$$\mathbf{M}\mathbf{a}_i = \nu_i \mathbf{a}_i , \quad i = 1, ..., N .$$

Further proceeding as in Appendix 2.A, we obtain

$$x_k(t) = v_{i_k}(t) ,$$

so that (2.66) gives

$$Q_{\mathcal{X}} = \sum_{k=1}^{N} Q_{v_{i_k}} = \sum_{k=1}^{N} \mu_{i_k} .$$

This is finally minimized by choosing the indices $i_k = k$, so that $Q_{\mathcal{X}} = \sum_{k=1}^{N} \mu_{i_k}$ becomes the sum of the N *smallest* eigenvalues. Hence, we obtain

$$x_k(t) = v_k(t), \quad k = 1, .., N \qquad \text{and} \qquad Q_{\mathcal{X}} = \sum_{k=1}^{N} \mu_k ,$$

which proves the first part of Theorem 2.2. The second part (pertaining to Case 2, i.e., maximization of $Q_{\mathcal{X}}$) is proved in an analogous manner.

Appendix 2.C: Proof of Theorem 2.3

According to Theorem 2.2, the optimum space is spanned by the first N eigenfunctions $v_k(t)$ of the restricted operator $\mathbf{H}_{\mathcal{C}}$. The eigenvalues μ_k and eigenfunctions $v_k(t)$ of $\mathbf{H}_{\mathcal{C}}$ are defined by the eigenequation

$$(\mathbf{H}_{\mathcal{C}} v_k)(t) = \mu_k v_k(t) . \tag{2.C.1}$$

Let us now assume that an orthonormal basis $\{c_j(t)\}_{j=1}^{N_C}$ of the constraint space $\mathcal{C}$ is available. The eigenfunctions $v_k(t)$ of $\mathbf{H}_\mathcal{C}$, being themselves elements of $\mathcal{C}$, can then be expanded as

$$v_k(t) \;=\; \sum_{j=1}^{N_C} \alpha_{kj}\, c_j(t) \qquad \text{with} \quad \alpha_{kj} = \langle v_k, c_j \rangle \,.$$

Inserting this expansion into the eigenequation (2.C.1), we obtain

$$\sum_{j=1}^{N_C} \alpha_{kj}\, \big(\mathbf{H}_\mathcal{C}\, c_j\big)(t) \;=\; \mu_k \sum_{j=1}^{N_C} \alpha_{kj}\, c_j(t) \,.$$

Taking the inner product with $c_i(t)$ of both sides and using the orthonormality of the $c_j(t)$ gives

$$\sum_{j=1}^{N_C} \alpha_{kj}\, \langle \mathbf{H}_\mathcal{C}\, c_j, c_i \rangle \;=\; \mu_k\, \alpha_{ki} \,. \tag{2.C.2}$$

Introducing the matrix $\boldsymbol{\Gamma}$ and the vectors $\mathbf{v}_k$ according to

$$(\boldsymbol{\Gamma})_{ij} \;=\; \langle \mathbf{H} c_j, c_i \rangle \;=\; \langle \mathbf{H}_\mathcal{C}\, c_j, c_i \rangle\,, \qquad (\mathbf{v}_k)_i \;=\; \alpha_{ki} \;=\; \langle v_k, c_i \rangle\,,$$

(2.C.2) can be compactly written as

$$\boldsymbol{\Gamma} \mathbf{v}_k \;=\; \mu_k \mathbf{v}_k \,,$$

which shows that the μ_k and $\mathbf{v}_k$ are the eigenvalues and eigenvectors, respectively, of the Hermitian matrix $\boldsymbol{\Gamma}$, as claimed in Theorem 2.3.

3 TIME-FREQUENCY LOCALIZATION OF LINEAR SIGNAL SPACES

Several basic properties of the WD of a linear signal space have been discussed in the previous chapter. In this chapter, we consider properties that are related to the *geometry* of the WD and to the *TF localization* of a space. For example, we study the relation between the orthogonality of two spaces and the TF disjointness of their WDs. We introduce a distinction between "simple" and "sophisticated" spaces that has a direct bearing on the shape of the WD and the TF shifts caused by the space's projection operator. Quantitative measures of a space's TF concentration are introduced, bounds for these concentration measures (generalizing conventional uncertainty relations) are formulated, and the "maximally concentrated" spaces attaining these bounds are explicitly characterized.

Furthermore, a quantity called "TF localization error" is introduced as a measure for the TF localization of a space in a given TF region. The minimization of this measure will be used in Chapter 4 as a mathematical basis for the TF synthesis of linear signal spaces, which in turn will allow the design of "TF projection filters" as discussed in Chapter 5.

This chapter is organized as follows. Section 3.1 discusses the shape of the WD and introduces the distinction between "simple" and "sophisticated" spaces. Section 3.2 investigates the relations existing between space affiliation and TF affiliation, and between space orthogonality and TF disjointness. Sections 3.3 and 3.4 introduce quantities measuring the TF concentration of a space as well as lower bounds on these concentration quantities that can be considered as "uncertainty relations for signal spaces." Finally, the localization of a space in a TF region is quantified in Section 3.5 via the introduction of a "TF localization error."

3.1 Shape of the Wigner Distribution

The fact that the WD of a space cannot be interpreted as a TF energy density in a strict (pointwise) manner grants the WD certain liberties regarding its detailed behavior or shape. This section considers various aspects of the local shape of the WD of a space.

3.1.1   Local Averages and Height Bounds

We continue our discussion of local WD averages

$$S_{\mathcal{X}}^{(h)}(t,f) = \left\| h_{\mathcal{X}}^{(t,f)} \right\|^2 = \langle W_{h^{(t,f)}}, W_{\mathcal{X}} \rangle$$

$$= \int_{t'} \int_{f'} W_h(t'-t, f'-f)\, W_{\mathcal{X}}(t', f')\, dt'\, df' \qquad (3.1)$$

commenced in Section 2.5 (see (2.48)). We recall that $h^{(t,f)}(t') = h(t'-t)\, e^{j2\pi ft'}$ denotes a TF shifted version of a basic "test function" $h(t')$. With our assumption that $\|h\|^2 = 1$, we have $\left\| h^{(t,f)} \right\|^2 = 1$ and therefore $0 \leq \left\| h_{\mathcal{X}}^{(t,f)} \right\|^2 \leq 1$. Inserting this into (3.1) yields

$$0 \leq \int_{t'} \int_{f'} W_h(t'-t, f'-f)\, W_{\mathcal{X}}(t', f')\, dt'\, df' \leq 1$$

$$\text{where} \quad \int_t \int_f W_h(t, f)\, dt\, df = 1\,.$$

This means that suitable local averages of the WD of a space may not be negative or larger than 1. The weighting function $W_h(t'-t, f'-f)$ may be concentrated about any TF point (t, f) inside a "Heisenberg ellipse" whose effective TF area is on the order of 1 [Claasen and Mecklenbräuker, 1984, de Bruijn, 1967, Janssen, 1997b]. Hence, *on any such TF Heisenberg ellipse, the WD of a space may not be consistently negative or consistently larger than 1*, although it may be negative or larger than 1 on smaller subregions of a TF Heisenberg ellipse.

We also recall from Section 2.3 (see (2.20)) that for a normalized signal $s(t)$

$$\int_t \int_f W_s(t,f)\, W_{\mathcal{X}}(t,f)\, dt\, df \;=\; \langle W_s, W_{\mathcal{X}} \rangle \;=\; \|s_{\mathcal{X}}\|^2$$

$$=\; \begin{cases} \|s\|^2 = 1 & \text{for } s(t) \in \mathcal{X} \\ \quad 0 & \text{for } s(t) \perp \mathcal{X} \,. \end{cases}$$

Thus, a weighted WD integral with weighting function $W_s(t,f)$ will be 1 if the (normalized) signal $s(t)$ is an element of the space $\mathcal{X}$, and 0 if $s(t)$ is orthogonal to $\mathcal{X}$. If we take the signal $s(t)$ to be well TF concentrated, then it follows that $W_{\mathcal{X}}(t,f)$, apart from local oscillations, must be approximately 1 (for $s(t) \in \mathcal{X}$) or 0 (for $s(t) \perp \mathcal{X}$) within the TF support of $s(t)$.

We emphasize that these results do not restrict the *pointwise* behavior of the WD of a space. In a given single point of the TF plane, the WD may be arbitrarily large and arbitrarily negative, as shown by the following example.

Example 3.1. Let us TF shift the even-signal space $\mathcal{E}$ (see Section 2.4) by time t_0 and frequency f_0 according to (2.33). With (2.39), the WD of the resulting space $\tilde{\mathcal{E}}$ is

$$W_{\tilde{\mathcal{E}}}(t,f) \;=\; \frac{1}{2} + \frac{1}{4}\,\delta(t - t_0)\,\delta(f - f_0)\,.$$

This WD is everywhere $1/2$, apart from the TF point (t_0, f_0) where it is infinite. Playing the same trick on the odd-signal space $\mathcal{O}$ yields (cf. (2.40))

$$W_{\tilde{\mathcal{O}}}(t,f) \;=\; \frac{1}{2} - \frac{1}{4}\,\delta(t - t_0)\,\delta(f - f_0)\,,$$

which is minus infinity at (t_0, f_0). $\qquad\qquad\square$

The spaces considered in this example have infinite dimension. In the case of a finite-dimensional space, the WD cannot be infinite. Specifically, the WD is bounded as

$$|W_{\mathcal{X}}(t,f)| \;\leq\; 2\,N_{\mathcal{X}}\,.$$

Proof. Expressing the WD of a signal as an inner product and using Schwarz' inequality, it is easy to show that the WD of a normalized signal $x(t)$ is bounded as $|W_x(t,f)| \leq 2$. With (2.7), we then obtain

$$|W_{\mathcal{X}}(t,f)| \;=\; \left| \sum_{k=1}^{N_{\mathcal{X}}} W_{x_k}(t,f) \right| \;\leq\; \sum_{k=1}^{N_{\mathcal{X}}} |W_{x_k}(t,f)| \;\leq\; \sum_{k=1}^{N_{\mathcal{X}}} 2 \;=\; 2\,N_{\mathcal{X}}\,. \qquad\square$$

Fig. 3.1 shows the WDs of finite-dimensional approximations to the spaces $\tilde{\mathcal{E}}$ and $\tilde{\mathcal{O}}$ considered in Example 3.1.

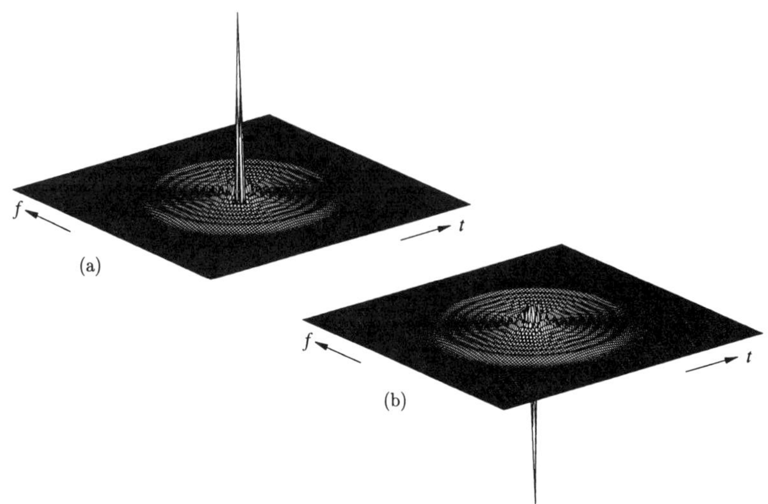

Figure 3.1. WDs of spaces with sharp positive or negative peaks: (a) WD of signal space spanned by the first 10 odd-indexed (i.e., even-symmetric) Hermite functions $h_k^{(T)}(t)$ ($k = 1, 3, .., 19$), (b) WD of signal space spanned by the first 10 even-indexed (i.e., odd-symmetric) Hermite functions $h_k^{(T)}(t)$ ($k = 2, 4, .., 20$). Inside their effective TF support and except for the central peak, the WDs shown oscillate about height $1/2$.

3.1.2 Sophisticated Spaces

Many spaces have WDs whose heights are approximately unity inside and approximately zero outside the space's effective TF support[1]. These spaces will be called *simple*. On the other hand, there exist spaces whose WDs contain substantial "interference terms" [Hlawatsch and Flandrin, 1997], i.e., structures which are oscillatory and involve large negative values. The next example considers a simple but fundamental special case.

Example 3.2. Let the one-dimensional space $\mathcal{X}$ be spanned by the single basis signal

$$x_1(t) = \frac{1}{\sqrt{2}} \left[y_1(t) + y_2(t) \right],$$

---

[1]Here, the term *approximately* allows larger deviations from zero or unity as long as these deviations are *local*, i.e., restricted to TF regions much smaller than a Heisenberg ellipse.

where $y_1(t)$ and $y_2(t)$ are two orthonormal signals which we assume to be located sufficiently far apart in the TF plane. Note that $\|x_1\| = 1$ due to the fact that $y_1(t)$ and $y_2(t)$ are orthonormal. The WD of the space $\mathcal{X}$ is

$$W_{\mathcal{X}}(t,f) \;=\; W_{x_1}(t,f) \;=\; \frac{1}{2}\,W_{y_1}(t,f) \;+\; \frac{1}{2}\,W_{y_2}(t,f) \;+\; \mathrm{Re}\{W_{y_1,y_2}(t,f)\}\,.$$

With our assumptions regarding $y_1(t)$ and $y_2(t)$, the "cross term" or "interference term" $\mathrm{Re}\{W_{y_1,y_2}(t,f)\}$ is an oscillatory structure involving positive and negative values to an equal extent, such that $\int_t \int_f \mathrm{Re}\{W_{y_1,y_2}(t,f)\}\,dt\,df = 0$ [Hlawatsch and Flandrin, 1997]. Typically, also the local averages considered in Sections 2.5 and 3.1.1 will be approximately zero. Hence, the oscillatory interference term does not contain any space energy. $\qquad\square$

This construction principle can be extended to more than just two signal components $y_k(t)$. Consider the basis signal

$$x_1(t) \;=\; \frac{1}{\sqrt{K}} \sum_{k=1}^{K} y_k(t)\,,$$

where the $y_k(t)$ are orthonormal and located sufficiently far apart in the TF plane. Here, $W_{\mathcal{X}}(t,f)$ consists of K "signal terms" $\frac{1}{K} W_{y_k}(t,f)$ and $K(K-1)/2$ oscillatory interference terms $\frac{2}{K} \mathrm{Re}\{W_{y_k,y_l}(t,f)\}$ (where $l > k$),

$$W_{\mathcal{X}}(t,f) \;=\; W_{x_1}(t,f) \;=\; \frac{1}{K} \sum_{k=1}^{K} W_{y_k}(t,f) \;+\; \frac{2}{K} \sum_{l=1}^{K} \sum_{k=1}^{l-1} \mathrm{Re}\{W_{y_k,y_l}(t,f)\}\,.$$

$$(3.2)$$

This construction can furthermore be extended to spaces with dimension larger than 1. The resulting spaces differ from the "simple" spaces as follows:

- The space's energy (and the space's WD) is spread over a TF region whose area is noticeably larger than the space's dimension, and the WD's height in this region is typically smaller than 1 (cf. the factor $1/K$ contained in the signal terms in (3.2)).

- The WD contains oscillatory interference terms that tend to involve large negative components and do not contain energy.

Spaces with this particular behavior will be called *sophisticated*. **Fig. 3.2** compares a simple one-dimensional space with a sophisticated one-dimensional space. A similar comparison is made in **Fig. 3.3** for dimension $N = 11$. Infinite-dimensional examples of sophisticated spaces are the spaces $\mathcal{E}$ and $\mathcal{O}$ introduced in Section 2.4 (see also Fig. 3.1).

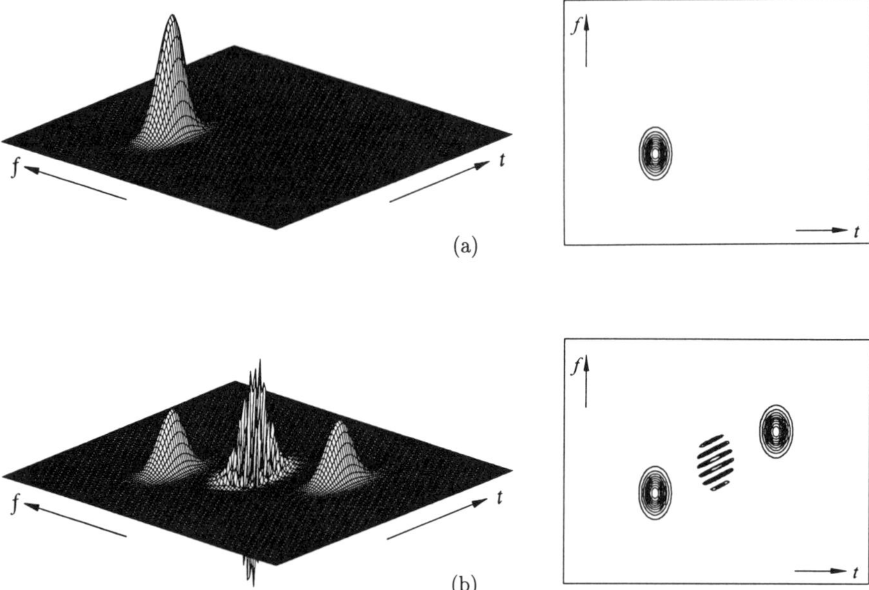

Figure 3.2. (a) WD of simple one-dimensional space, (b) WD of sophisticated one-dimensional space.

Apart from the shape of its WD, a sophisticated space has another interesting property. Let us consider the sophisticated space $\mathcal{X}$ from Example 3.2. Using the fact that $y_1(t)$ and $y_2(t)$ are orthonormal, the orthogonal projection of $y_1(t)$ onto $\mathcal{X}$ is obtained as

$$y_{1,\mathcal{X}}(t) \;=\; \langle y_1, x_1 \rangle \, x_1(t) \;=\; \frac{1}{\sqrt{2}}\, x_1(t) \;=\; \frac{1}{2}\left[y_1(t) + y_2(t)\right]. \tag{3.3}$$

We see that half of the energy of the projected signal is assigned to $y_1(t)$ (the input signal) but the other half is assigned to $y_2(t)$, which was assumed to be located far away from $y_1(t)$ in the TF plane. Thus, as a result of projecting $y_1(t)$ onto $\mathcal{X}$, a considerable part of signal energy is transferred from the TF location of the input signal $y_1(t)$ to the TF location of $y_2(t)$. We conclude that the orthogonal projection of a signal onto a sophisticated space may cause a partial (but noticeable) *TF displacement* of signal energy. Such a TF displacement will not be observed when projecting a signal onto a simple space. This *TF displacement effect* associated with sophisticated spaces will be further studied in the context of the *ambiguity function of a signal space* (see Chapter 7).

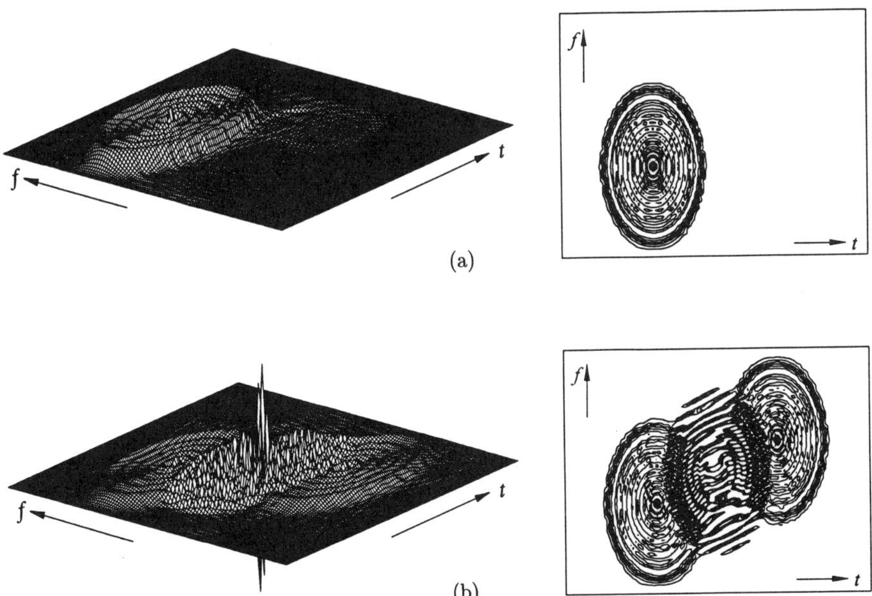

Figure 3.3. (a) WD of simple 11-dimensional space, (b) WD of sophisticated 11-dimensional space.

Specifically, we shall show in Section 7.4.1 that the ambiguity function of a signal space allows a more formal definition of sophisticated spaces.

3.2 Time-Frequency Disjointness and Time-Frequency Affiliation

The Moyal-type relations (2.16) and (2.20) can be used to derive some interesting geometric interpretations. Specifically, we can show that two spaces whose TF supports do not overlap are orthogonal, and that the TF support of a space that is a subspace of another space lies within the TF support of that other space [Hlawatsch and Kozek, 1993]. The same is true if one of the spaces is formally replaced by a signal.

3.2.1 Affiliation and Orthogonality

For later use in subsequent sections, we first introduce quantities measuring the relation of a space to another space, and of a signal to a space.

Two spaces. A space $\mathcal{Y}$ may be a subspace of another space $\mathcal{X}$, i.e., $\mathcal{Y} \subseteq \mathcal{X}$ which means that any $y(t) \in \mathcal{Y}$ is also a member of $\mathcal{X}$. Equivalently, we can say that "$\mathcal{Y}$ is *affiliated* to $\mathcal{X}$." The opposite case is that of *orthogonal* spaces, i.e., $\mathcal{Y} \perp \mathcal{X}$ which means that any $y(t) \in \mathcal{Y}$ is orthogonal to any $x(t) \in \mathcal{X}$.

Between these two extremes, intermediate situations exist. In particular, $\mathcal{Y}$ may be "nearly a subspace of $\mathcal{X}$" in the sense that $y_{\mathcal{X}}(t) \approx y(t)$ for any signal $y(t) \in \mathcal{Y}$. Also, $\mathcal{Y}$ may be "nearly orthogonal to $\mathcal{X}$" in the sense that $y_{\mathcal{X}}(t) \approx 0$ for any $y(t) \in \mathcal{Y}$. These vague notions can be made mathematically precise by introducing the *affiliation of a space $\mathcal{Y}$ to a space $\mathcal{X}$* as

$$a(\mathcal{Y}|\mathcal{X}) \triangleq \frac{N_{\mathcal{X},\mathcal{Y}}}{N_{\mathcal{Y}}}$$

where

$$N_{\mathcal{X},\mathcal{Y}} \triangleq \sum_{k=1}^{N_{\mathcal{X}}} \|x_{k,\mathcal{Y}}\|^2 = \sum_{l=1}^{N_{\mathcal{Y}}} \|y_{l,\mathcal{X}}\|^2 = \sum_{k=1}^{N_{\mathcal{X}}} \sum_{l=1}^{N_{\mathcal{Y}}} |\langle x_k, y_l \rangle|^2 .$$

The affiliation of $\mathcal{Y}$ to $\mathcal{X}$ can be written as

$$a(\mathcal{Y}|\mathcal{X}) = \frac{\sum_{l=1}^{N_{\mathcal{Y}}} \|y_{l,\mathcal{X}}\|^2}{\sum_{l=1}^{N_{\mathcal{Y}}} \|y_l\|^2} .$$

Since $N_{\mathcal{X},\mathcal{Y}} = \langle W_{\mathcal{X}}, W_{\mathcal{Y}} \rangle$ due to (2.16), the affiliation $a(\mathcal{Y}|\mathcal{X})$ can also be expressed in terms of the WDs of the spaces $\mathcal{X}$ and $\mathcal{Y}$ as

$$a(\mathcal{Y}|\mathcal{X}) = \frac{\langle W_{\mathcal{X}}, W_{\mathcal{Y}} \rangle}{N_{\mathcal{Y}}} = \frac{\int_t \int_f W_{\mathcal{X}}(t,f)\, W_{\mathcal{Y}}(t,f)\, dt\, df}{\int_t \int_f W_{\mathcal{Y}}(t,f)\, dt\, df} . \tag{3.4}$$

The following properties are easily checked (cf. (2.17)–(2.19)):

$$a(\mathcal{X}|\mathcal{Y}) = \frac{N_{\mathcal{Y}}}{N_{\mathcal{X}}} a(\mathcal{Y}|\mathcal{X}) ,$$

$$0 \leq a(\mathcal{Y}|\mathcal{X}) \leq \min\left\{ \frac{N_{\mathcal{X}}}{N_{\mathcal{Y}}}, 1 \right\} (\leq 1) ,$$

$$a(\mathcal{Y}|\mathcal{X}) = 0 \quad \Longleftrightarrow \quad \mathcal{Y} \perp \mathcal{X}$$
$$a(\mathcal{Y}|\mathcal{X}) = 1 \quad \Longleftrightarrow \quad \mathcal{Y} \subseteq \mathcal{X}$$
$$a(\mathcal{Y}|\mathcal{X}) = \frac{N_{\mathcal{X}}}{N_{\mathcal{Y}}} \quad \Longleftrightarrow \quad \mathcal{X} \subseteq \mathcal{Y} .$$

We shall say that "$\mathcal{Y}$ is nearly orthogonal to $\mathcal{X}$" if $a(\mathcal{Y}|\mathcal{X}) \approx 0$, and that "$\mathcal{Y}$ is nearly a subspace of $\mathcal{X}$" or "$\mathcal{Y}$ is nearly affiliated to $\mathcal{X}$" if $a(\mathcal{Y}|\mathcal{X}) \approx 1$.

Signal and space. The above discussion can be reformulated for a signal and a space. Let us introduce the *affiliation of a signal $s(t)$ to a space $\mathcal{X}$* as

$$a(s|\mathcal{X}) \triangleq \frac{\|s_{\mathcal{X}}\|^2}{\|s\|^2} \,.$$

Since $\|s_{\mathcal{X}}\|^2 = \langle W_{\mathcal{X}}, W_s \rangle$ due to (2.20), the affiliation $a(s|\mathcal{X})$ can be expressed in terms of the WDs of the signal $s(t)$ and the space $\mathcal{X}$ as

$$a(s|\mathcal{X}) = \frac{\langle W_{\mathcal{X}}, W_s \rangle}{E_s} = \frac{\int_t \int_f W_{\mathcal{X}}(t, f)\, W_s(t, f)\, dt\, df}{\int_t \int_f W_s(t, f)\, dt\, df} \,. \tag{3.5}$$

It is evident that (cf. (2.21)–(2.23))

$$0 \leq a(s|\mathcal{X}) \leq 1$$

with

$$a(s|\mathcal{X}) = 0 \qquad \Longleftrightarrow \qquad s \perp \mathcal{X}$$
$$a(s|\mathcal{X}) = 1 \qquad \Longleftrightarrow \qquad s \in \mathcal{X} \,.$$

Hence, "$s(t)$ is nearly orthogonal to $\mathcal{X}$" if $a(s|\mathcal{X}) \approx 0$, and "$s(t)$ is nearly an element of (or nearly affiliated to) $\mathcal{X}$" if $a(s|\mathcal{X}) \approx 1$.

3.2.2   Orthogonality and Time-Frequency Disjointness

We next study the relations between the TF disjointness and orthogonality of two spaces, or a signal and a space. We will see that the distinction between simple and sophisticated spaces plays an important role for these relations.

Two spaces. Let us consider two spaces $\mathcal{X}$ and $\mathcal{Y}$ that are *TF disjoint* in the sense that their TF supports (i.e., their WDs) do not overlap, which means

$$W_{\mathcal{X}}(t, f)\, W_{\mathcal{Y}}(t, f) = 0 \qquad \text{for all } t, f \,.$$

It follows that $\langle W_{\mathcal{X}}, W_{\mathcal{Y}} \rangle = 0$ which, due to (2.18), implies that $\mathcal{X}$ and $\mathcal{Y}$ are orthogonal. Hence, *two TF disjoint spaces are orthogonal.*

Since exact TF disjointness is a very restrictive condition, we note that this statement still holds in an approximate sense if the spaces are "nearly TF disjoint" in the sense that their *effective* TF supports do not overlap or, equivalently, $W_{\mathcal{X}}(t, f)\, W_{\mathcal{Y}}(t, f) \approx 0$. Indeed, it follows from (3.4) that

$$W_{\mathcal{X}}(t, f)\, W_{\mathcal{Y}}(t, f) \approx 0 \quad \text{for all } t, f \qquad \Rightarrow \qquad a(\mathcal{Y}|\mathcal{X}) \approx 0 \,.$$

Hence, nearly TF disjoint spaces are nearly orthogonal in the sense of Section 3.2.1. Two effectively TF disjoint spaces are depicted schematically in part (a) of **Fig. 3.4.**

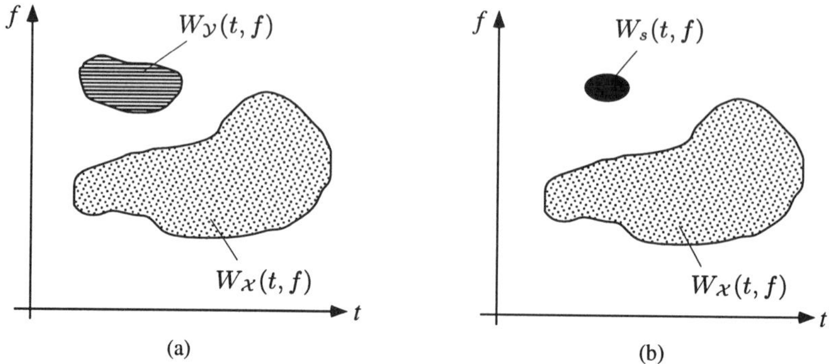

Figure 3.4.   Effective TF disjointness: (a) of two spaces, (b) of a signal and a space.

Although two (effectively) TF disjoint spaces are (nearly) orthogonal, the converse need not be true: *Two orthogonal spaces need not be TF disjoint.* The next example contrasts two fundamentally different cases using trivial one-dimensional spaces.

Example 3.3. Consider two orthonormal signals $x(t)$ and $y(t)$ which are TF disjoint in the sense that their WDs do not overlap. Let $\mathcal{X}$ and $\mathcal{Y}$ be the one-dimensional signal spaces spanned by $x(t)$ and $y(t)$, respectively. The spaces $\mathcal{X}$ and $\mathcal{Y}$ are obviously orthogonal. Furthermore, since $W_{\mathcal{X}}(t,f) = W_x(t,f)$ and $W_{\mathcal{Y}}(t,f) = W_y(t,f)$, and since $x(t)$ and $y(t)$ are TF disjoint, the spaces $\mathcal{X}$ and $\mathcal{Y}$ are TF disjoint as well.

Next, define

$$a(t) \,=\, \frac{1}{\sqrt{2}}\left[x(t) + y(t)\right], \qquad b(t) \,=\, \frac{1}{\sqrt{2}}\left[x(t) - y(t)\right],$$

and let $\mathcal{A}$ and $\mathcal{B}$ be the one-dimensional signal spaces spanned by $a(t)$ and $b(t)$, respectively. The signals $a(t)$ and $b(t)$ are easily shown to be orthonormal, and thus the spaces $\mathcal{A}$ and $\mathcal{B}$ are orthogonal. The spaces' WDs are

$$W_{\mathcal{A}}(t,f) \,=\, W_a(t,f) \,=\, \frac{1}{2}\,W_x(t,f) + \frac{1}{2}\,W_y(t,f) + \mathrm{Re}\{W_{x,y}(t,f)\}\,,$$

$$W_{\mathcal{B}}(t,f) \,=\, W_b(t,f) \,=\, \frac{1}{2}\,W_x(t,f) + \frac{1}{2}\,W_y(t,f) - \mathrm{Re}\{W_{x,y}(t,f)\}\,;$$

they are identical apart from the sign of the interference term $\mathrm{Re}\{W_{x,y}(t,f)\}$. Hence, the spaces $\mathcal{A}$ and $\mathcal{B}$ are not TF disjoint; rather, they completely overlap in the TF plane. We have thus constructed spaces that are orthogonal without being TF disjoint. Note that $\mathcal{A}$ and $\mathcal{B}$ are *sophisticated* spaces as defined

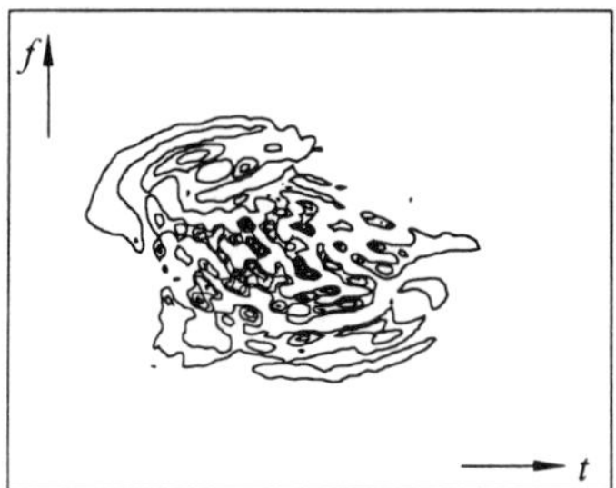
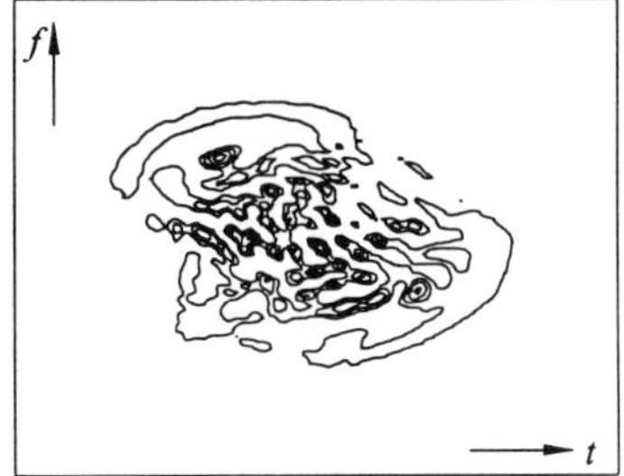

Figure 3.5. WDs of two orthogonal spaces (both with dimension $N = 8$) whose effective TF supports overlap.

in Section 3.1.2, and that the spaces' sophistication is necessary for having complete TF overlap and orthogonality simultaneously.   □

This construction principle can easily be generalized to spaces with dimensions larger than 1. Consider two orthonormal signal sets $\{x_k(t)\}_{k=1}^K$ and $\{y_l(t)\}_{l=1}^L$ with the property that all pairs of signals $x_k(t)$ and $y_l(t)$ are orthonormal and TF disjoint for any k and l. Then, the spaces

$$\mathcal{X} = \mathrm{span}\{x_k(t)\}_{k=1}^K \,, \qquad \mathcal{Y} = \mathrm{span}\{y_l(t)\}_{l=1}^L$$

are both orthogonal and TF disjoint. On the other hand, we can construct two new orthonormal signal sets $\{a_m(t)\}_{m=1}^M$ and $\{b_n(t)\}_{n=1}^N$ where the $a_m(t)$ and $b_n(t)$ are linear combinations of signals taken from $\{x_k(t)\}_{k=1}^K \cup \{y_l(t)\}_{l=1}^L$, and choose the coefficients of these linear combinations such that $a_m(t)$ and $b_n(t)$ are orthonormal for any m and n. Then, the spaces

$$\mathcal{A} = \mathrm{span}\{a_m(t)\}_{m=1}^M \,, \qquad \mathcal{B} = \mathrm{span}\{b_n(t)\}_{n=1}^N$$

are orthogonal but *not* TF disjoint in general. An example of orthogonal but not TF disjoint spaces is shown in **Fig. 3.5**. Another (infinite-dimensional) example is given by the spaces $\mathcal{E}$ and $\mathcal{O}$ from Section 2.4 (see also Fig. 3.1).

Signal and space. A similar discussion applies to the relation between a space and a signal. Consider a signal $s(t)$ and a space $\mathcal{X}$ that are TF disjoint in the sense that their effective TF supports (or WDs) do not overlap, i.e.,

$$W_s(t, f)\, W_{\mathcal{X}}(t, f) = 0 \qquad \text{for all } t, f \,,$$

as shown schematically in Fig. 3.4(b). It follows that $\langle W_s, W_{\mathcal{X}} \rangle = 0$ which, due to (2.22), implies that the signal $s(t)$ is orthogonal to the space $\mathcal{X}$. Hence, *a signal that is TF disjoint with respect to a space is orthogonal to the space.* This can be extended to the practically more important case of approximate TF disjointness: *A signal that is nearly TF disjoint with respect to a space is nearly orthogonal to the space,* where "nearly orthogonal" stands for $a(s|\mathcal{X}) \approx 0$ (cf. (3.5)). Again, the converse statement is not generally valid: *A signal may be orthogonal to a space without being TF disjoint with respect to that space.* We shall again illustrate two basic situations by means of a simple example.

Example 3.4. Consider a space spanned by an orthonormal basis $\{x_k(t)\}_{k=1}^{N_\mathcal{X}}$,

$$\mathcal{X} = \text{span}\{x_k(t)\}_{k=1}^{N_\mathcal{X}} \ .$$

Furthermore take a signal $s(t)$ that is orthogonal to, and TF disjoint with respect to, all basis signals $x_k(t)$. Then, the signal $s(t)$ and the space $\mathcal{X}$ are easily seen to be both orthogonal and TF disjoint.

On the other hand, we can construct a new signal $\tilde{s}(t)$ and a new first basis signal $\tilde{x}_1(t)$ according to

$$\tilde{s}(t) = C \frac{1}{\sqrt{2}} \left[s_n(t) + x_1(t)\right] , \qquad \tilde{x}_1(t) = \frac{1}{\sqrt{2}} \left[s_n(t) - x_1(t)\right] ,$$

where C is an arbitrary complex factor and $s_n(t) = s(t)/\|s\|$ denotes the normalized signal $s(t)$. It is easily checked that $\tilde{s}(t)$ and $\tilde{x}_1(t)$ are orthogonal, and further, that $\tilde{x}_1(t)$ is orthogonal to all basis signals $x_k(t)$ with $k \neq 1$. Let $\tilde{\mathcal{X}}$ be the space spanned by the basis $\{x_k(t)\}_{k=1}^{N_\mathcal{X}}$ with $x_1(t)$ replaced by $\tilde{x}_1(t)$,

$$\tilde{\mathcal{X}} = \text{span}\{\tilde{x}_1(t), x_2(t), x_3(t), ..., x_{N_\mathcal{X}}(t)\} \ .$$

We note that $\{\tilde{x}_1(t), x_2(t), x_3(t), ..., x_{N_\mathcal{X}}(t)\}$ is an orthonormal basis of $\tilde{\mathcal{X}}$. The signal $\tilde{s}(t)$ is again orthogonal to the new space $\tilde{\mathcal{X}}$. However, the WDs of $\tilde{s}(t)$ and $\tilde{\mathcal{X}}$ are

$$W_{\tilde{s}}(t, f) = |C|^2 \left[\frac{1}{2} W_{s_n}(t, f) + \frac{1}{2} W_{x_1}(t, f) + \text{Re}\{W_{s_n, x_1}(t, f)\}\right]$$

$$W_{\tilde{\mathcal{X}}}(t, f) = W_{\tilde{x}_1}(t, f) + \sum_{k=2}^{N_\mathcal{X}} W_{x_k}(t, f)$$

$$= \frac{1}{2} W_{s_n}(t, f) + \frac{1}{2} W_{x_1}(t, f) - \text{Re}\{W_{s_n, x_1}(t, f)\} + \sum_{k=2}^{N_\mathcal{X}} W_{x_k}(t, f) ,$$

from which it follows that the effective TF support of $\tilde{s}(t)$ is contained in the effective TF support of $\tilde{\mathcal{X}}$ and, hence, that $\tilde{s}(t)$ and $\tilde{\mathcal{X}}$ are not TF disjoint. Note that the space $\tilde{\mathcal{X}}$ is (partly) sophisticated due to its first basis signal $\tilde{x}_1(t)$. An example of this construction is depicted in **Fig. 3.6.**   □

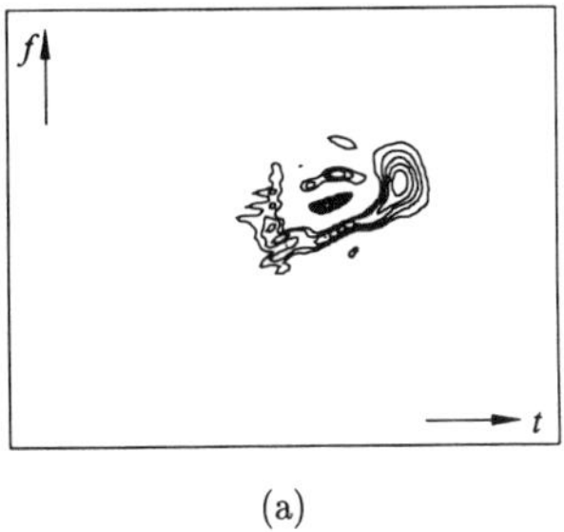
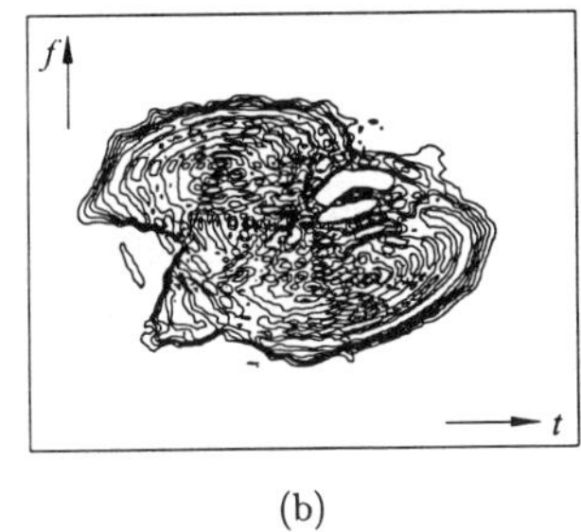

(a)   (b)

Figure 3.6.   WD of a signal (a) and WD of a space (b). The signal is orthogonal to the space but not TF disjoint with respect to the space.

3.2.3 Affiliation and Time-Frequency Affiliation

Just as TF disjointness is related to orthogonality, the opposite situation—where the TF support of a space is entirely contained within the TF support of another space—is related to the case where one space is a subspace of (or affiliated to) the other space. Again, an analogous relation exists if the first space is replaced by a signal.

Two spaces. Let us say that a space $\mathcal{Y}$ is *TF affiliated* to a space $\mathcal{X}$ if the entire effective TF support of $\mathcal{Y}$ is contained in the effective TF support of $\mathcal{X}$. This situation is depicted schematically in part (a) of **Fig. 3.7**.

Consider now a space $\mathcal{Y}$ that is a subspace of another space $\mathcal{X}$, i.e., $\mathcal{Y} \subseteq \mathcal{X}$. From (2.19), we know that $\langle W_{\mathcal{X}}, W_{\mathcal{Y}} \rangle = N_{\mathcal{Y}}$, which can also be written as

$$\int_t \int_f W_{\mathcal{Y}}(t,f)\, W_{\mathcal{X}}(t,f)\, dt\, df \;=\; \int_t \int_f W_{\mathcal{Y}}(t,f)\, dt\, df \;.$$

Since, according to Section 3.1.1, $W_{\mathcal{X}}(t,f)$ may not be consistently larger than 1, this implies that the effective TF support of $\mathcal{Y}$ must be inside the effective TF support of $\mathcal{X}$. Hence, *if $\mathcal{Y} \subseteq \mathcal{X}$, then $\mathcal{Y}$ is TF affiliated to $\mathcal{X}$*, or in other words, *affiliation entails TF affiliation*. The converse is however not true in general: *A space that is TF affiliated to another space need not be a subspace of that space*. In fact, the basic construction used in Examples 3.3 and 3.4 can be employed to construct two spaces that are *orthogonal* even though one space is TF affiliated to the other, as sketched in the following example.

Example 3.5. Consider two orthonormal signal sets $\{x_k(t)\}_{k=1}^{K}$ and $\{y_l(t)\}_{l=1}^{L}$ with the property that all pairs of signals $x_k(t)$ and $y_l(t)$ are orthonormal and

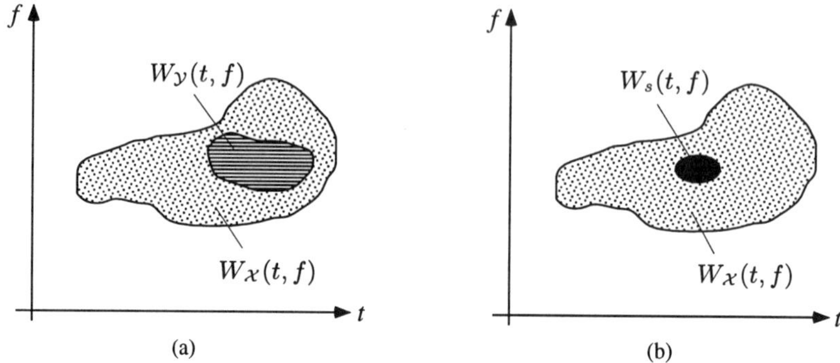

Figure 3.7.   TF affiliation: (a) the space $\mathcal{Y}$ is TF affiliated to the space $\mathcal{X}$, (b) the signal $s(t)$ is TF affiliated to the space $\mathcal{X}$.

TF disjoint for any k and l. Suppose that $L \leq K$. We derive two signal sets $\{\tilde{x}_k(t)\}_{k=1}^{L}$ and $\{\tilde{y}_l(t)\}_{l=1}^{L}$ according to

$$\tilde{x}_k(t) = \frac{1}{\sqrt{2}}\left[x_k(t) + y_k(t)\right], \qquad \tilde{y}_k(t) = \frac{1}{\sqrt{2}}\left[x_k(t) - y_k(t)\right], \qquad k = 1, ..., L.$$

These signal sets are again orthonormal, and also the signals $\tilde{x}_k(t)$ and $\tilde{y}_l(t)$ are orthonormal for any k and l. Define $\tilde{\mathcal{X}}$ to be the space spanned by the orthonormal basis $\{\tilde{x}_k(t)\}_{k=1}^{L} \cup \{x_k(t)\}_{k=L+1}^{K}$ (i.e., the first L original basis signals $x_k(t)$ are replaced by the new basis signals $\tilde{x}_k(t)$). Furthermore define $\tilde{\mathcal{Y}}$ to be the space spanned by the orthonormal basis $\{\tilde{y}_k(t)\}_{k=1}^{L}$. Thus, we have

$$\tilde{\mathcal{X}} = \text{span}\{\tilde{x}_1(t), ..., \tilde{x}_L(t), x_{L+1}(t), ..., x_K(t)\}, \qquad \tilde{\mathcal{Y}} = \text{span}\{\tilde{y}_1(t), ..., \tilde{y}_L(t)\}.$$

Then, the spaces $\tilde{\mathcal{X}}$ and $\tilde{\mathcal{Y}}$ are orthogonal, and yet $\tilde{\mathcal{Y}}$ is TF affiliated to $\tilde{\mathcal{X}}$. Note that the spaces $\tilde{\mathcal{X}}$ and $\tilde{\mathcal{Y}}$ are both sophisticated. An example (with $K = L = 8$) has been given in Fig. 3.5. $\qquad\qquad\qquad\qquad\qquad\qquad\qquad\quad$ $\square$

Signal and space. A signal $s(t)$ will be said to be *TF affiliated* to a space $\mathcal{X}$ if the entire effective TF support of $s(t)$ is contained in the effective TF support of $\mathcal{X}$, as depicted schematically in Fig. 3.7(b). This property of TF affiliation can be related to the situation where the signal is an element of (or affiliated to) the space.

Assume that $s(t)$ is an element of $\mathcal{X}$, i.e., $s(t) \in \mathcal{X}$. From (2.23), we know that $\langle W_s, W_{\mathcal{X}} \rangle = E_s$, which can also be written as

$$\int_t \int_f W_s(t, f)\, W_{\mathcal{X}}(t, f)\, dt\, df = \int_t \int_f W_s(t, f)\, dt\, df.$$

Since $W_{\mathcal{X}}(t,f)$ may not be consistently >1, this implies that the effective TF support of $s(t)$ must be inside that of $\mathcal{X}$. Thus, *if $s(t) \in \mathcal{X}$, then $s(t)$ is TF affiliated to $\mathcal{X}$*, or in other words, **affiliation entails TF affiliation**. Again, the converse is not true in general: *A signal may be TF affiliated to a space without being an element of this space.* In fact, in Example 3.4 we have constructed a signal and a space such that the signal is TF affiliated to the space and yet orthogonal to the space. An example has been shown in Fig. 3.6.

3.3 Uncertainty Relations

The *uncertainty principle* [Papoulis, 1984b, de Bruijn, 1967, Folland and Sitaram, 1997] states that a signal cannot be arbitrarily concentrated with respect to both time and frequency. Equivalently, a signal's effective TF support has a certain minimum extension. In this section, a classical formulation of the uncertainty principle will be generalized to linear signal spaces [Hlawatsch and Kozek, 1991]. The resulting *uncertainty relation for linear signal spaces* provides a lower bound on the joint TF concentration of any signal space, and also shows that the *Hermite spaces* introduced in Section 2.4 have maximum TF concentration. An alternative formulation of the uncertainty principle that applies to band-limited signals and spaces will be considered in Section 3.4.

3.3.1 Time-Frequency Moments

One simple way of expressing the "TF uncertainty" of a signal $x(t)$ involves the signal's second-order moments. Specifically, the *root-mean-square duration* T_x and the *root-mean-square bandwidth* F_x are defined by

$$
T_x^2 \triangleq \frac{m_x^{(2)}}{E_x} = \frac{\int_t t^2\, d_x(t)\, dt}{\int_t d_x(t)\, dt} = \frac{\int_t \int_f t^2\, W_x(t,f)\, dt\, df}{\int_t \int_f W_x(t,f)\, dt\, df}
$$

$$
F_x^2 \triangleq \frac{M_x^{(2)}}{E_x} = \frac{\int_f f^2\, D_x(f)\, df}{\int_f D_x(f)\, df} = \frac{\int_t \int_f f^2\, W_x(t,f)\, dt\, df}{\int_t \int_f W_x(t,f)\, dt\, df} \; ,
$$

where $m_x^{(2)}$ and $M_x^{(2)}$ are the second-order moments considered in (2.9) and (2.10), respectively. A related quantity is the *TF radius* $\kappa_x^{(T)}$ defined by [de Bruijn, 1967, Claasen and Mecklenbräuker, 1980, Janssen, 1982]

$$
\left[\kappa_x^{(T)}\right]^2 \triangleq \frac{\int_t \int_f \left[\left(\frac{t}{T}\right)^2 + (Tf)^2\right] W_x(t,f)\, dt\, df}{\int_t \int_f W_x(t,f)\, dt\, df} = \left(\frac{T_x}{T}\right)^2 + (TF_x)^2 , \quad (3.6)
$$

where $T > 0$ is a fixed reference (normalization) time parameter.

Using the quadratic space representations $E_\mathcal{X}$, $d_\mathcal{X}(t)$, $D_\mathcal{X}(f)$, $m_\mathcal{X}^{(2)}$, and $M_\mathcal{X}^{(2)}$ defined in Section 2.2.2, it is straightforward to reformulate the quantities T_x, F_x, and $\kappa_x^{(T)}$ for a linear signal space $\mathcal{X}$. We define the *root-mean-square duration* $T_\mathcal{X}$, the *root-mean-square bandwidth* $F_\mathcal{X}$, and the *TF radius* $\kappa_\mathcal{X}^{(T)}$ of a signal space $\mathcal{X}$ as

$$
T_\mathcal{X}^2 \;\triangleq\; \frac{m_\mathcal{X}^{(2)}}{E_\mathcal{X}} \;=\; \frac{\int_t t^2\, d_\mathcal{X}(t)\, dt}{\int_t d_\mathcal{X}(t)\, dt} \;=\; \frac{\int_t \int_f t^2\, W_\mathcal{X}(t,f)\, dt\, df}{\int_t \int_f W_\mathcal{X}(t,f)\, dt\, df}
\tag{3.7}
$$

$$
F_\mathcal{X}^2 \;\triangleq\; \frac{M_\mathcal{X}^{(2)}}{E_\mathcal{X}} \;=\; \frac{\int_f f^2\, D_\mathcal{X}(f)\, df}{\int_f D_\mathcal{X}(f)\, df} \;=\; \frac{\int_t \int_f f^2\, W_\mathcal{X}(t,f)\, dt\, df}{\int_t \int_f W_\mathcal{X}(t,f)\, dt\, df}
\tag{3.8}
$$

$$
\left[\kappa_\mathcal{X}^{(T)}\right]^2 \;\triangleq\; \frac{\int_t \int_f \left[\left(\frac{t}{T}\right)^2 + (Tf)^2\right] W_\mathcal{X}(t,f)\, dt\, df}{\int_t \int_f W_\mathcal{X}(t,f)\, dt\, df} \;=\; \left(\frac{T_\mathcal{X}}{T}\right)^2 + (T F_\mathcal{X})^2 .
$$

It is easily seen that the squared space spreads $T_\mathcal{X}^2$, $F_\mathcal{X}^2$, and $\left[\kappa_\mathcal{X}^{(T)}\right]^2$ are the averages of the respective squared spreads of all orthonormal basis signals $x_k(t)$,

$$
T_\mathcal{X}^2 = \frac{1}{N_\mathcal{X}} \sum_{k=1}^{N_\mathcal{X}} T_{x_k}^2 , \qquad F_\mathcal{X}^2 = \frac{1}{N_\mathcal{X}} \sum_{k=1}^{N_\mathcal{X}} F_{x_k}^2 , \qquad \left[\kappa_\mathcal{X}^{(T)}\right]^2 = \frac{1}{N_\mathcal{X}} \sum_{k=1}^{N_\mathcal{X}} \left[\kappa_{x_k}^{(T)}\right]^2 .
\tag{3.9}
$$

3.3.2 Uncertainty Relations

The classical *uncertainty relation for signals* places a lower bound on the "duration-bandwidth product" $T_x F_x$ [Papoulis, 1984b, de Bruijn, 1967, Folland and Sitaram, 1997]: for any signal $x(t)$,

$$
T_x F_x \;\geq\; \frac{1}{4\pi} .
\tag{3.10}
$$

The lower bound is attained by a Gaussian signal or, equivalently, the first Hermite signal $h_1^{(T)}(t) = \sqrt{\sqrt{2}/T}\, e^{-\pi(t/T)^2}$ (see Section 2.4),

$$
T_x F_x = \frac{1}{4\pi} \qquad \Longleftrightarrow \qquad x(t) = C\, h_1^{(T)}(t)
$$

where $T > 0$ is arbitrary and C is an arbitrary constant factor. An equivalent formulation of this uncertainty relation is the following lower bound on the TF radius $\kappa_x^{(T)}$: for any signal $x(t)$ and any reference time T,

$$
\kappa_x^{(T)} \;\geq\; \sqrt{\frac{1}{2\pi}} .
\tag{3.11}
$$

The lower bound is again attained by the Gaussian signal $h_1^{(T)}(t)$,

$$\kappa_x^{(T)} = \sqrt{\frac{1}{2\pi}} \qquad \Longleftrightarrow \qquad x(t) = C\, h_1^{(T)}(t) \, .$$

Here, the time parameter T of the Gaussian signal $h_1^{(T)}(t)$ has to be equal to the reference time used in the TF radius $\kappa_x^{(T)}$.

This uncertainty relation for signals can be extended to linear signal spaces, as is stated in the following theorem whose proof is given in Appendix 3.A.

Theorem 3.1 (Uncertainty Relation for Linear Signal Spaces) *The "duration-bandwidth product" $T_\mathcal{X} F_\mathcal{X}$ of a linear signal space $\mathcal{X}$ with dimension N is bounded from below as*

$$T_\mathcal{X} F_\mathcal{X} \;\geq\; \frac{N}{4\pi} \, . \tag{3.12}$$

The lower bound is attained if and only if $\mathcal{X}$ is the Hermite space of dimension N,

$$T_\mathcal{X} F_\mathcal{X} = \frac{N}{4\pi} \;\; for \; N_\mathcal{X} = N \qquad \Longleftrightarrow \qquad \mathcal{X} = \mathcal{H}_N^{(T)} \, ,$$

where T is arbitrary.

It is shown in Appendix 3.A that an equivalent formulation of this uncertainty relation is the following lower bound on the TF radius $\kappa_\mathcal{X}^{(T)}$: for any N-dimensional space $\mathcal{X}$ and any reference time T,

$$\kappa_\mathcal{X}^{(T)} \;\geq\; \sqrt{\frac{N}{2\pi}} \, . \tag{3.13}$$

Again, the lower bound is attained by the Hermite space of dimension N,

$$\kappa_\mathcal{X}^{(T)} = \sqrt{\frac{N}{2\pi}} \;\; for \; N_\mathcal{X} = N \qquad \Longleftrightarrow \qquad \mathcal{X} = \mathcal{H}_N^{(T)} \, ,$$

where the time parameter T of $\mathcal{H}_N^{(T)}$ equals the reference time used in $\kappa_\mathcal{X}^{(T)}$.

The above results show that the Hermite spaces $\mathcal{H}_N^{(T)}$ (see Section 2.4) are the spaces with minimum overall spread in the TF plane, assuming that we measure TF spread by the duration-bandwidth product $T_\mathcal{X} F_\mathcal{X}$ or, equivalently, by the TF radius $\kappa_\mathcal{X}^{(T)}$. Note that the classical uncertainty relations for signals, (3.10) and (3.11), are a special case of the uncertainty relations for signal spaces, (3.12) and (3.13); they are re-obtained for dimension $N = 1$. In the "signal case," the minimum TF uncertainty is attained by a Gaussian signal which is equivalent to a Hermite space of dimension 1.

3.4 Uncertainty Relation for Band-limited Spaces

A further manifestation of the uncertainty principle is the fact that a signal that is strictly band-limited may not be strictly time-limited. This section reviews a classical quantitative statement of this principle and provides an extension to linear signal spaces. It is shown that the *prolate spheroidal spaces* introduced in Section 2.4 are the band-limited spaces with maximum time concentration.

3.4.1 Temporal Concentration

Besides the root-mean-square duration T_x considered in Section 3.3, an alternative quantity characterizing the temporal concentration of a signal $x(t)$ is the fraction of energy assumed within a prescribed time interval $[-T/2, T/2]$ of length T [Papoulis, 1984b],

$$\alpha_x^{(T)} \triangleq \frac{E_x^{(T)}}{E_x} \qquad \text{with} \quad E_x^{(T)} = \int_{-T/2}^{T/2} d_x(t)\, dt \,.$$

We shall call $\alpha_x^{(T)}$ the *temporal concentration* of the signal $x(t)$ in the interval $[-T/2, T/2]$. The temporal concentration is bounded as

$$0 \le \alpha_x^{(T)} \le 1 \,,$$

with $\alpha_x^{(T)} = 1$ for a signal exactly time-limited to $[-T/2, T/2]$ and $\alpha_x^{(T)} = 0$ for a signal located entirely outside $[-T/2, T/2]$. We note that the temporal concentration can be expressed in terms of the WD as

$$\alpha_x^{(T)} = \frac{\int_{-T/2}^{T/2} \int_f W_x(t, f)\, dt\, df}{\int_t \int_f W_x(t, f)\, dt\, df} \,.$$

We now extend the temporal concentration $\alpha_x^{(T)}$ to linear signal spaces. We define the *temporal concentration of a linear signal space* $\mathcal{X}$ as

$$\alpha_{\mathcal{X}}^{(T)} \triangleq \frac{E_{\mathcal{X}}^{(T)}}{N_{\mathcal{X}}} \qquad \text{with} \quad E_{\mathcal{X}}^{(T)} = \int_{-T/2}^{T/2} d_{\mathcal{X}}(t)\, dt \,,$$

where $d_{\mathcal{X}}(t)$ is the temporal energy density (instantaneous power) of $\mathcal{X}$ (cf. Section 2.2.2). The temporal concentration of a space $\mathcal{X}$ is the average of the temporal concentrations of all orthonormal basis signals $x_k(t)$,

$$\alpha_{\mathcal{X}}^{(T)} = \frac{1}{N_{\mathcal{X}}} \sum_{k=1}^{N_{\mathcal{X}}} \alpha_{x_k}^{(T)} \,.$$

It is again bounded as

$$0 \leq \alpha_{\mathcal{X}}^{(T)} \leq 1 \, ,$$

with $\alpha_{\mathcal{X}}^{(T)} = 1$ for a space whose instantaneous power $d_{\mathcal{X}}(t)$ is exactly time-limited to $[-T/2, T/2]$ and $\alpha_{\mathcal{X}}^{(T)} = 0$ for a space whose instantaneous power is located entirely outside $[-T/2, T/2]$. The temporal concentration of a space $\mathcal{X}$ can be expressed in terms of the WD of $\mathcal{X}$ according to

$$\alpha_{\mathcal{X}}^{(T)} \;=\; \frac{\int_{-T/2}^{T/2} \int_f W_{\mathcal{X}}(t, f) \, dt \, df}{\int_t \int_f W_{\mathcal{X}}(t, f) \, dt \, df} \; . \tag{3.14}$$

3.4.2 *Uncertainty Relation for Band-limited Spaces*

We now consider a signal $x(t)$ that is exactly band-limited to a frequency band $[-F/2, F/2]$. It is then well known that $x(t)$ cannot be exactly time-limited to any finite interval $[-T/2, T/2]$, i.e., $\alpha_x^{(T)} \neq 1$ for any $T < \infty$. This raises the question as to which signal band-limited to $[-F/2, F/2]$ has maximum temporal concentration $\alpha_x^{(T)}$ for given interval length T, and how large this maximum concentration is.

The answer to this question involves the family of prolate spheroidal wave functions $p_k^{(T,F)}(t)$ [Slepian and Pollak, 1961, Landau and Pollak, 1961, Papoulis, 1984b] previously considered in Section 2.4. We recall that the prolate spheroidal wave functions $p_k^{(T,F)}(t)$ are the normalized eigenfunctions of the linear operator $\mathbf{K}^{(T,F)}$ obtained by composing the time-limitation operator on $[-T/2, T/2]$ with the band-limitation operator (idealized lowpass filter) on $[-F/2, F/2]$ (see (2.47)) The eigenvalues of $\mathbf{K}^{(T,F)}$, $\lambda_k^{(T,F)}$, are real-valued with $0 < \lambda_k^{(T,F)} < 1$, and the eigenfunctions $p_k^{(T,F)}(t)$ constitute an orthonormal basis of $\mathcal{F}[-F/2, F/2]$, the linear space of all signals band-limited to $[-F/2, F/2]$ [Papoulis, 1984b]. Thus, any signal band-limited to $[-F/2, F/2]$ can be expanded into the prolate spheroidal wave functions $p_k^{(T,F)}(t)$. In what follows, we assume that the eigenvalues $\lambda_k^{(T,F)}$ are arranged in decreasing order, i.e., $1 > \lambda_1^{(T,F)} > \lambda_2^{(T,F)} > ... > 0$, which induces a corresponding ordering of the eigenfunctions $p_k^{(T,F)}(t)$.

A classical result [Slepian and Pollak, 1961, Landau and Pollak, 1961, Papoulis, 1984b] states that the signal band-limited to $[-F/2, F/2]$ which has maximum temporal concentration $\alpha_x^{(T)}$ is, within a factor, the first prolate spheroidal wave function $p_1^{(T,F)}(t)$, and the corresponding (maximum) temporal concentration is the first (i.e., maximum) eigenvalue $\lambda_1^{(T,F)}$. This can be restated as follows: for any signal $x(t)$ bandlimited to $[-F/2, F/2]$, the temporal concentration $\alpha_x^{(T)}$ is bounded from above as

$$\alpha_x^{(T)} \leq \lambda_1^{(T,F)} \quad (< 1) \,,$$

where the upper bound is attained by the first prolate spheroidal wave function,

$$\alpha_x^{(T)} = \lambda_1^{(T,F)} \qquad \Longleftrightarrow \qquad x(t) = C\, p_1^{(T,F)}(t) \,.$$

This can be considered as an uncertainty relation for band-limited signals. A difference from the uncertainty relation considered in Section 3.3.2 is the fact that the concentration bound and the signal attaining it are not given by an explicit expression but rather as the solution to an integral equation.

The uncertainty relation for band-limited signals can be extended to linear signal spaces, as is stated in the next theorem whose proof is provided in Appendix 3.B.

Theorem 3.2 (Uncertainty Relation for Band-limited Linear Signal Spaces) *The temporal concentration $\alpha_{\mathcal{X}}^{(T)}$ of any N-dimensional linear signal space $\mathcal{X} \subseteq \mathcal{F}[-F/2, F/2]$ band-limited to $[-F/2, F/2]$ is bounded from above as*

$$\alpha_{\mathcal{X}}^{(T)} \leq \frac{1}{N} \sum_{k=1}^{N} \lambda_k^{(T,F)}, \qquad (3.15)$$

where the $\lambda_k^{(T,F)}$ $(k = 1, .., N)$ are the N largest eigenvalues of the operator $\mathbf{K}^{(T,F)}$. The upper bound is attained if and only if $\mathcal{X}$ is the prolate spheroidal space of dimension N with parameters T and F,

$$\alpha_{\mathcal{X}}^{(T)} = \frac{1}{N} \sum_{k=1}^{N} \lambda_k^{(T,F)} \quad \text{for } \mathcal{X} \subseteq \mathcal{F}[-F/2, F/2] \text{ and } N_{\mathcal{X}} = N$$

$$\Longleftrightarrow \quad \mathcal{X} = \mathcal{P}_N^{(T,F)} \,.$$

This result shows that the prolate spheroidal spaces $\mathcal{P}_N^{(T,F)}$ (see Section 2.4) are the band-limited spaces with maximum time concentration, assuming that we measure time concentration by $\alpha_{\mathcal{X}}^{(T)}$. Note that the conventional uncertainty relation for band-limited signals is a special case of the uncertainty relation for band-limited signal spaces; it is reobtained for dimension $N = 1$.

It is furthermore clear that an analogous results holds if the roles of time and frequency are interchanged: among all N-dimensional signal spaces $\mathcal{X} \subseteq \mathcal{T}[-T/2, T/2]$ exactly time-limited to an interval $[-T/2, T/2]$, the space maximizing the *spectral concentration*

$$\tilde{\alpha}_{\mathcal{X}}^{(F)} \triangleq \frac{\tilde{E}_{\mathcal{X}}^{(F)}}{N_{\mathcal{X}}} \qquad \text{with} \quad \tilde{E}_{\mathcal{X}}^{(F)} = \int_{-F/2}^{F/2} D_{\mathcal{X}}(f)\, df$$

is a "dual prolate spheroidal space" $\tilde{\mathcal{P}}_N^{(F,T)}$ spanned by orthonormal basis signals $\tilde{p}_k^{(F,T)}(t)$ whose Fourier transforms $\tilde{P}_k^{(F,T)}(f)$ are the first N prolate spheroidal wave functions (with T and F interchanged),

$$\tilde{P}_k^{(F,T)}(f) = p_k^{(F,T)}(f), \qquad k = 1, .., N.$$

3.5 Localization in a Time-Frequency Region

We have seen in Section 2.4 that there exist some specific *infinite-dimensional* spaces whose WD is exactly unity inside and zero outside a TF region with *infinite* area. The next theorem states that this precise behavior will never be observed in the case of a *bounded* TF region (i.e., a region that is entirely contained in a rectangle with finite sides).

Theorem 3.3 *The WD of a linear signal space cannot be identically zero outside a bounded TF region.*

Proof. Suppose the WD were exactly zero outside a bounded TF region. Then, it follows from the marginal properties (2.26), (2.27) that the instantaneous power $d_\mathcal{X}(t)$ and the spectral energy density $D_\mathcal{X}(f)$ must be exactly zero outside a finite time interval and a finite frequency interval, respectively. Since $d_\mathcal{X}(t) = \sum_{k=1}^{N_\mathcal{X}} |x_k(t)|^2$ and $D_\mathcal{X}(f) = \sum_{k=1}^{N_\mathcal{X}} |X_k(f)|^2$, this implies that the space's basis signals $x_k(t)$ have simultaneously finite time support and finite frequency support. But this cannot be true since a signal may not be simultaneously time-limited and band-limited (cf. Section 3.4.2). $\quad\square$

Although Theorem 3.3 prohibits the existence of a space with bounded TF support, it is yet true that the *effective* TF support of any *finite*-dimensional space $\mathcal{X} \subseteq \mathcal{L}_2(\mathbb{R})$ is bounded. We shall introduce two related quantities, the *regional TF concentration* and the *TF localization error*, for measuring the extent to which a space is localized in a given TF region. These quantities will play an important role for the TF synthesis of signal spaces (see Chapter 4).

3.5.1 Regional Time-Frequency Concentration

We first ask how well a linear signal space is concentrated in a given TF region R, i.e., what part of the space's energy is contained in R. Since the total energy of a linear signal space $\mathcal{X}$ equals the integral of $W_\mathcal{X}(t, f)$ over the entire TF plane (see (2.25)), it is natural to define the space's "energy content in R" as[2]

---

[2]An analogous quantity has been considered for a *signal* in [Flandrin, 1988] and [Janssen, 1997b]. We emphasize that $E_\mathcal{X}^{(R)}$ is not guaranteed to be positive or smaller than the total energy $N_\mathcal{X}$.

the integral of $W_{\mathcal{X}}(t, f)$ over the TF region R,

$$E_{\mathcal{X}}^{(R)} \triangleq \iint_{(t,f) \in R} W_{\mathcal{X}}(t, f) \, dt \, df \, .$$

Introducing the indicator function (or "characteristic function") of the TF region R

$$I_R(t, f) \triangleq \begin{cases} 1, & (t, f) \in R \\ 0, & (t, f) \notin R \, , \end{cases}$$

the regional energy can be expressed as a weighted WD integral in the sense of Section 2.7 (see (2.57)),

$$E_{\mathcal{X}}^{(R)} = \langle W_{\mathcal{X}}, I_R \rangle = \int_t \int_f W_{\mathcal{X}}(t, f) \, I_R(t, f) \, dt \, df \, .$$

Furthermore, we define the *regional TF concentration* $\rho(\mathcal{X}, R)$ of a space $\mathcal{X}$ in the TF region R as the ratio of the regional energy $E_{\mathcal{X}}^{(R)}$ and the total energy (dimension) $E_{\mathcal{X}} = N_{\mathcal{X}}$,

$$\rho(\mathcal{X}, R) \triangleq \frac{E_{\mathcal{X}}^{(R)}}{N_{\mathcal{X}}} = \frac{\iint_{(t,f) \in R} W_{\mathcal{X}}(t, f) \, dt \, df}{\int_t \int_f W_{\mathcal{X}}(t, f) \, dt \, df} = \frac{1}{N_{\mathcal{X}}} \langle W_{\mathcal{X}}, I_R \rangle \, . \tag{3.16}$$

An upper bound on the regional TF concentration $\rho(\mathcal{X}, R)$ involving the space's dimension $N_{\mathcal{X}}$ and the area of the TF region R,

$$A_R \triangleq \iint_{(t,f) \in R} dt \, df = \int_t \int_f I_R(t, f) \, dt \, df = \int_t \int_f [I_R(t, f)]^2 \, dt \, df \, , \tag{3.17}$$

is provided by the *concentration inequality* stated in the following theorem [Hlawatsch and Kozek, 1994].

Theorem 3.4 (Concentration Inequality) *The regional TF concentration is bounded from above as*

$$|\rho(\mathcal{X}, R)| \leq \sqrt{\frac{A_R}{N_{\mathcal{X}}}} \, . \tag{3.18}$$

Proof. Using $\|W_{\mathcal{X}}\|^2 = \int_t \int_f [W_{\mathcal{X}}(t, f)]^2 \, dt \, df = N_{\mathcal{X}}$ (see (2.13)) and $\|I_R\|^2 = \int_t \int_f [I_R(t, f)]^2 \, dt \, df = A_R$, we have by Schwarz' inequality

$$|\rho(\mathcal{X}, R)| = \frac{|\langle W_{\mathcal{X}}, I_R \rangle|}{N_{\mathcal{X}}} \leq \frac{\|W_{\mathcal{X}}\| \, \|I_R\|}{N_{\mathcal{X}}} = \frac{\sqrt{N_{\mathcal{X}}} \sqrt{A_R}}{N_{\mathcal{X}}} = \sqrt{\frac{A_R}{N_{\mathcal{X}}}} \, . \qquad \square$$

From the concentration inequality, it follows that *good regional TF concentration of a space (i.e., $\rho(\mathcal{X}, R) \approx 1$) requires that the space's dimension $N_\mathcal{X}$ is not larger than the TF region's area A_R.* In fact, as $N_\mathcal{X}$ grows, the space's TF support grows as well; for $N_\mathcal{X} > A_R$, the space will "spill over" the TF region R and the regional concentration will be poor. We conclude that *the area of the effective TF support of a space is at least the space's dimension.*

3.5.2 Time-Frequency Localization Error

The regional TF concentration of a space in a TF region is sensitive to a "spilling-over" of space energy outside the region. However, it does not give an indication as to how well the region is "covered energetically" by the space. In fact, the space's WD may be (nearly) zero over wide parts of the region and the regional TF concentration may still be close to one. We now look for a quantity which is sensitive to both a spilling-over of space energy outside the region *and* energy gaps inside the region. Let us consider a hypothetical "idealized space" whose WD is unity inside the given TF region R and zero outside R, i.e., the space's WD equals the indicator function $I_R(t, f)$ of the region R. From Theorem 3.3, we know that such an idealized space does not exist for a bounded region R. In order to measure the actual deviation from this idealized situation, we introduce the *TF localization error $\epsilon(\mathcal{X}, R)$* of the space $\mathcal{X}$ with respect to the TF region R as

$$\epsilon^2(\mathcal{X}, R) \triangleq \iint_{(t,f) \in R} \left[1 - W_\mathcal{X}(t, f) \right]^2 dt\, df + \iint_{(t,f) \notin R} \left[0 - W_\mathcal{X}(t, f) \right]^2 dt\, df.$$

Using the indicator function $I_R(t, f)$, the squared TF localization error can be compactly written as

$$\epsilon^2(\mathcal{X}, R) = \| I_R - W_\mathcal{X} \|^2 = \int_t \int_f \left[I_R(t, f) - W_\mathcal{X}(t, f) \right]^2 dt\, df. \tag{3.19}$$

It is easily shown that the regional TF concentration $\rho(\mathcal{X}, R)$ and the TF localization error $\epsilon(\mathcal{X}, R)$ are related as

$$\epsilon^2(\mathcal{X}, R) = A_R + N_\mathcal{X} \left[1 - 2\rho(\mathcal{X}, R) \right]. \tag{3.20}$$

The following lower bound on the TF localization error involves the space's dimension $N_\mathcal{X}$ and the TF region's area A_R [Hlawatsch and Kozek, 1994]:

Corollary 3.1 (Localization Inequality) *The TF localization error is bounded from below as*

$$\epsilon(\mathcal{X}, R) \geq \left| \sqrt{N_\mathcal{X}} - \sqrt{A_R} \right|. \tag{3.21}$$

Proof. Insert (3.20) into the concentration inequality (3.18).   □

This *localization inequality* shows that *good localization of a space in a TF region (i.e., small TF localization error) is possible only if the space's dimension is approximately equal to the region's area.* Thus, if the TF localization error of a space with respect to the space's effective TF support is small, then we may expect that *the TF area of the effective TF support of a space is approximately equal to the space's dimension.* This will not be true for a *sophisticated space* as defined in Section 3.1.2. Indeed, due to the shape of the WD of a sophisticated space, the sophisticated TF localization error of a sophisticated space with respect to the space's effective TF support will never be small. However, in the case of a *simple* space we may indeed expect that the area of its TF support is approximately equal to the space's dimension.

3.5.3 Affiliation Inequality

There exists an interesting inequality involving the affiliation introduced in Section 3.2.1, the regional TF concentration, and the TF localization error.

Two spaces. If a space $\mathcal{X}$ is well localized in a TF region R, i.e., the TF localization error $\epsilon(\mathcal{X}, R)$ is small, and if a second space $\mathcal{Y}$ is well concentrated in R, i.e., the regional TF concentration $\rho(\mathcal{Y}, R)$ is nearly one, then we would expect[3] that $\mathcal{Y}$ is "nearly a subspace of $\mathcal{X}$" in the sense that the affiliation $a(\mathcal{Y}|\mathcal{X})$ is nearly one. This situation is depicted in part (a) of **Fig. 3.8**.

On the other hand, if $\mathcal{X}$ is again well localized in a TF region R, i.e., the TF localization error $\epsilon(\mathcal{X}, R)$ is small, but now the second space $\mathcal{Y}$ is well outside R, i.e., the regional TF concentration $\rho(\mathcal{Y}, R)$ is nearly zero, then we would expect that $\mathcal{Y}$ is "nearly orthogonal to $\mathcal{X}$" in the sense that the affiliation $a(\mathcal{Y}|\mathcal{X})$ is nearly zero. This situation is shown in Fig. 3.8(b).

The following inequality [Hlawatsch and Kozek, 1994] provides lower and upper bounds on the affiliation $a(\mathcal{Y}|\mathcal{X})$ in terms of $\epsilon(\mathcal{X}, R)$ and $\rho(\mathcal{Y}, R)$.

Theorem 3.5 (Affiliation Inequality) *The affiliation of $\mathcal{Y}$ to $\mathcal{X}$ is bounded as*

$$\rho(\mathcal{Y}, R) - \frac{\epsilon(\mathcal{X}, R)}{\sqrt{N_y}} \leq a(\mathcal{Y}|\mathcal{X}) \leq \rho(\mathcal{Y}, R) + \frac{\epsilon(\mathcal{X}, R)}{\sqrt{N_y}} , \qquad (3.22)$$

where R is an arbitrary TF region.

---

[3]If $\epsilon(\mathcal{X}, R)$ is small and $\rho(\mathcal{Y}, R)$ is nearly one, then $\mathcal{Y}$ will also be TF affiliated to $\mathcal{X}$. In general, the fact that $\mathcal{Y}$ is TF affiliated to $\mathcal{X}$ does not guarantee that $\mathcal{Y}$ is affiliated to $\mathcal{X}$—see the counterexamples (sophisticated spaces) discussed in Section 3.2.3. However, in our case the small TF localization error $\epsilon(\mathcal{X}, R)$ does not allow $\mathcal{X}$ to be a sophisticated space.

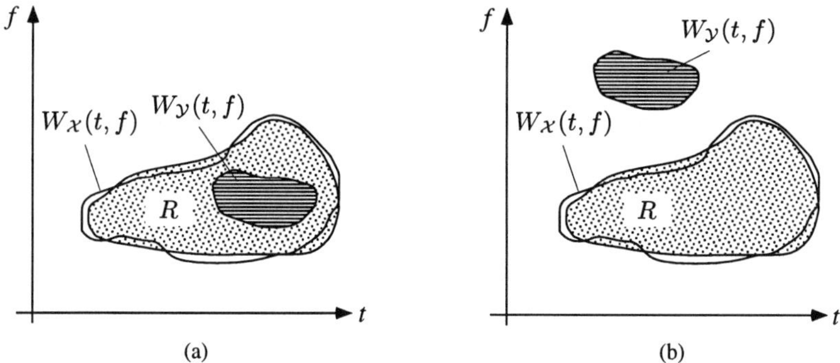

Figure 3.8. Discussion of the affiliation inequality (3.22): (a) If $\mathcal{X}$ is well localized in R and $\mathcal{Y}$ is well concentrated in R, then $\mathcal{Y}$ is approximately affiliated to $\mathcal{X}$. (b) If $\mathcal{X}$ is well localized in R but $\mathcal{Y}$ is outside R, then $\mathcal{Y}$ is approximately orthogonal to $\mathcal{X}$.

Proof. Using Schwarz' inequality, we have

$$
\left| a(\mathcal{Y}|\mathcal{X}) - \rho(\mathcal{Y},R) \right| = \left| \frac{\langle W_{\mathcal{Y}}, W_{\mathcal{X}} \rangle}{N_{\mathcal{Y}}} - \frac{\langle W_{\mathcal{Y}}, I_R \rangle}{N_{\mathcal{Y}}} \right| = \frac{\left| \langle W_{\mathcal{Y}}, W_{\mathcal{X}} - I_R \rangle \right|}{N_{\mathcal{Y}}}
$$

$$
\leq \frac{\|W_{\mathcal{Y}}\| \, \|W_{\mathcal{X}} - I_R\|}{N_{\mathcal{Y}}} = \frac{\sqrt{N_{\mathcal{Y}}} \, \epsilon(\mathcal{X},R)}{N_{\mathcal{Y}}} = \frac{\epsilon(\mathcal{X},R)}{\sqrt{N_{\mathcal{Y}}}} . \qquad \Box
$$

The affiliation inequality can be simplified in two cases. If the effective TF support of $\mathcal{Y}$ is well inside R, then $\rho(\mathcal{Y},R) \approx 1$ and we obtain approximately

$$
1 - \frac{\epsilon(\mathcal{X},R)}{\sqrt{N_{\mathcal{Y}}}} \leq a(\mathcal{Y}|\mathcal{X}) \leq 1 ,
$$

where we used $a(\mathcal{Y}|\mathcal{X}) \leq 1$. On the other hand, if the effective TF support of $\mathcal{Y}$ is well outside R, then $\rho(\mathcal{Y},R) \approx 0$ and we obtain approximately

$$
0 \leq a(\mathcal{Y}|\mathcal{X}) \leq \frac{\epsilon(\mathcal{X},R)}{\sqrt{N_{\mathcal{Y}}}} ,
$$

where we used $0 \leq a(\mathcal{Y}|\mathcal{X})$. In either case, the maximum deviation from the expected result $a(\mathcal{Y}|\mathcal{X}) = 1$ or $a(\mathcal{Y}|\mathcal{X}) = 0$ is proportional to the TF localization error $\epsilon(\mathcal{X},R)$.

Signal and space. The above discussion can be reformulated for a signal and a space. In analogy to the regional TF concentration of a space, we define the regional TF concentration $\rho(s, R)$ of a signal $s(t)$ in a TF region R as the ratio of the "regional energy" $E_s^{(R)} = \iint_{(t,f)\in R} W_s(t, f)\, dt\, df$ and the total energy E_s [Flandrin, 1988],

$$\rho(s, R) \;\triangleq\; \frac{E_s^{(R)}}{E_s} \;=\; \frac{\iint_{(t,f)\in R} W_s(t, f)\, dt\, df}{\int_t \int_f W_s(t, f)\, dt\, df} \;=\; \frac{\langle W_s, I_R \rangle}{E_s}\,.$$

Then, the affiliation of $s(t)$ to $\mathcal{X}$, the regional TF concentration of $s(t)$ in the TF region R, and the TF localization error of $\mathcal{X}$ with respect to this TF region R are related by the inequality

$$\rho(s, R) - \epsilon(\mathcal{X}, R) \;\leq\; a(s|\mathcal{X}) \;\leq\; \rho(s, R) + \epsilon(\mathcal{X}, R)\,,$$

which is of course just a simple special case (with $N_{\mathcal{Y}} = 1$) of the affiliation inequality (3.22). Again, this inequality can be simplified in two cases. If $s(t)$ is well inside R, then $\rho(s, R) \approx 1$ and we obtain approximately

$$1 - \epsilon(\mathcal{X}, R) \;\leq\; a(s|\mathcal{X}) \;\leq\; 1\,.$$

If $s(t)$ is well outside R, then $\rho(s, R) \approx 0$ and we have approximately

$$0 \;\leq\; a(s|\mathcal{X}) \;\leq\; \epsilon(\mathcal{X}, R)\,.$$

However, since $\epsilon(\mathcal{X}, R)$ is typically on the order of 1 or even larger, these bounds on $a(s|\mathcal{X})$ are quite loose.

Appendix 3.A: Proof of Theorem 3.1

We first review some facts about the Hermite functions $h_k^{(T)}(t)$ (see (2.43)) spanning the Hermite spaces $\mathcal{H}_N^{(T)}$. The $h_k^{(T)}(t)$ are the eigenfunctions of the linear operator $\mathbf{L}^{(T)}$ defined as[4]

$$\left(\mathbf{L}^{(T)} x\right)(t) \;=\; \left(\frac{t}{T}\right)^2 x(t) - \left(\frac{T}{2\pi}\right)^2 \frac{d^2}{dt^2}\, x(t)\,,$$

so that $\left(\mathbf{L}^{(T)} h_k^{(T)}\right)(t) \;=\; \lambda_k\, h_k^{(T)}(t)$ with the eigenvalues $\lambda_k \;=\; \frac{1}{2\pi}(2k - 1)$

---

[4]This follows from the fact that the Hermite functions $h_k^{(T)}(t)$ can be written as scaled versions $h_k^{(T)}(t) = \frac{1}{\sqrt{T}} h_k\left(\frac{t}{T}\right)$ of the functions

$$h_k(\alpha) \;\triangleq\; C_{k-1}\, H_{k-1}\left(\sqrt{2\pi}\,\alpha\right) e^{-\pi\alpha^2}\,, \quad k = 1, 2, \dots$$

which are well known [Wilcox, 1991] to be the eigenfunctions of the operator $\mathbf{L}$ defined as

$$(\mathbf{L}x)(\alpha) \;=\; \alpha^2 x(\alpha) - \frac{1}{(2\pi)^2} \frac{d^2}{d\alpha^2}\, x(\alpha)\,,$$

with eigenvalues $\lambda_k = \frac{1}{2\pi}(2k - 1)$.

($k = 1, 2, ...$). The operator $\mathbf{L}^{(T)}$ is closely related to the TF radius $\kappa_x^{(T)}$. Indeed, the squared TF radius $\left[\kappa_x^{(T)}\right]^2$ in (3.6) can be written as a normalized version of a "TF moment" $\mu_x^{(T)}$,

$$\left[\kappa_x^{(T)}\right]^2 = \frac{\mu_x^{(T)}}{E_x} \tag{3.A.1}$$

with

$$\mu_x^{(T)} = \int_t \int_f \left[\left(\frac{t}{T}\right)^2 + (Tf)^2\right] W_x(t,f)\, dt\, df = \frac{m_x^{(2)}}{T^2} + T^2 M_x^{(2)}$$

(cf. (2.9), (2.10)), and it easily checked that this TF moment is the quadratic form generated by the operator $\mathbf{L}^{(T)}$,

$$\mu_x^{(T)} = \left\langle \mathbf{L}^{(T)} x, x \right\rangle .$$

Furthermore, the eigenvalues λ_k of $\mathbf{L}^{(T)}$ equal the TF moments of the Hermite functions $h_k^{(T)}(t)$,

$$\left[\kappa_x^{(T)}\right]^2\bigg|_{x(t)=h_k^{(T)}(t)} = \mu_x^{(T)}\bigg|_{x(t)=h_k^{(T)}(t)} = \left\langle \mathbf{L}^{(T)} h_k^{(T)}, h_k^{(T)} \right\rangle = \lambda_k = \frac{1}{2\pi}(2k-1)$$

(note that here $\left[\kappa_x^{(T)}\right]^2 = \mu_x^{(T)}$ since the Hermite functions are normalized). This relation can be generalized as

$$\left\langle \mathbf{L}^{(T)} h_k^{(T)}, h_l^{(T)} \right\rangle = \lambda_k\, \delta_{kl} .$$

We shall now show the uncertainty relation for signal spaces, Eq. (3.12). We do this by first showing the uncertainty relation (3.13) and then proving that (3.12) and (3.13) are equivalent. Relation (3.13) will be proved by direct minimization of the TF radius $\kappa_{\mathcal{X}}^{(T)}$ for fixed space dimension $N_{\mathcal{X}} = N$.

In analogy to (3.A.1), the squared TF radius of a space $\mathcal{X}$ can be written as a normalized version of a "TF moment" $\mu_{\mathcal{X}}^{(T)}$,

$$\left[\kappa_{\mathcal{X}}^{(T)}\right]^2 = \frac{\mu_{\mathcal{X}}^{(T)}}{N_{\mathcal{X}}}$$

with

$$\mu_{\mathcal{X}}^{(T)} = \int_t \int_f \left[\left(\frac{t}{T}\right)^2 + (Tf)^2\right] W_{\mathcal{X}}(t,f)\, dt\, df = \frac{m_{\mathcal{X}}^{(2)}}{T^2} + T^2 M_{\mathcal{X}}^{(2)} ,$$

where

$$\mu_{\mathcal{X}}^{(T)} = \sum_{k=1}^{N_{\mathcal{X}}} \mu_{x_k}^{(T)} = \sum_{k=1}^{N_{\mathcal{X}}} \left\langle \mathbf{L}^{(T)} x_k, x_k \right\rangle . \tag{3.A.2}$$

We now look for the space $\mathcal{X}$ (with given dimension $N_\mathcal{X} = N$) minimizing the TF radius $\kappa_\mathcal{X}^{(T)}$. Since the dimension $N_\mathcal{X}$ is fixed, an equivalent problem is the minimization of the TF moment $\mu_\mathcal{X}^{(T)}$, i.e., the optimum space is

$$\mathcal{X}_{\mathrm{opt}}(N) \;=\; \arg \min_{N_\mathcal{X}=N} \mu_\mathcal{X}^{(T)} \,.$$

With (3.A.2), the results of Theorem 2.1 (see Section 2.7.1) are directly applicable to our minimization problem. Hence, $\mathcal{X}_{\mathrm{opt}}(N)$ is spanned by the first N eigenfunctions of the operator $\mathbf{L}^{(T)}$, i.e., the first N Hermite functions, so that $\mathcal{X}_{\mathrm{opt}}(N)$ is the Hermite space of dimension N:

$$\mathcal{X}_{\mathrm{opt}}(N) \;=\; \mathrm{span}\{h_k^{(T)}(t)\}_{k=1}^{N} \;=\; \mathcal{H}_N^{(T)} \,,$$

as claimed in Theorem 3.1. Furthermore, the minimum TF moment becomes (see Theorem 2.1)

$$\mu_\mathcal{X}^{(T)}\Big|_{\min} \;=\; \sum_{k=1}^{N} \lambda_k \;=\; \frac{1}{2\pi} \sum_{k=1}^{N} (2k-1) \;=\; \frac{N^2}{2\pi}\,,$$

from which the minimum TF radius is finally obtained as

$$\kappa_\mathcal{X}^{(T)}\Big|_{\min} \;=\; \kappa_\mathcal{X}^{(T)}\Big|_{\mathcal{X}=\mathcal{H}_N^{(T)}} \;=\; \sqrt{\frac{1}{N}\,\mu_\mathcal{X}^{(T)}\Big|_{\min}} \;=\; \sqrt{\frac{1}{N}\,\frac{N^2}{2\pi}} \;=\; \sqrt{\frac{N}{2\pi}}\,,$$

which is the lower bound (3.13) to be proved.

It remains to show that the result just proved is equivalent to the uncertainty relation (3.12), $T_\mathcal{X} F_\mathcal{X} \geq \frac{N}{4\pi}$. Let $\mathcal{X}$ be a space with root-mean-square duration $T_\mathcal{X}$ and root-mean-square bandwidth $F_\mathcal{X}$. For given $T_\mathcal{X}$ and $F_\mathcal{X}$, we can always find a reference time parameter $T = T_0$ such that[5]

$$T_\mathcal{X} F_\mathcal{X} \;=\; \frac{1}{2}\big[\kappa_\mathcal{X}^{(T_0)}\big]^2 \,.$$

Then, since $\big[\kappa_\mathcal{X}^{(T_0)}\big]^2 \geq \frac{N}{2\pi}$ according to (3.13), we obtain the bound (3.12),

$$T_\mathcal{X} F_\mathcal{X} \;\geq\; \frac{1}{2}\frac{N}{2\pi} \;=\; \frac{N}{4\pi}\,.$$

---

[5] To see this, consider the quantity

$$\left(\frac{T_\mathcal{X}}{T} - TF_\mathcal{X}\right)^2 \;=\; \left(\frac{T_\mathcal{X}}{T}\right)^2 + (TF_\mathcal{X})^2 - 2T_\mathcal{X} F_\mathcal{X} \;=\; \big[\kappa_\mathcal{X}^{(T)}\big]^2 - 2T_\mathcal{X} F_\mathcal{X}\,.$$

For $T = T_0 \overset{\triangle}{=} \sqrt{T_\mathcal{X}/F_\mathcal{X}}$, there is $\left(\frac{T_\mathcal{X}}{T} - TF_\mathcal{X}\right)^2 = 0$ and hence $\big[\kappa_\mathcal{X}^{(T)}\big]^2 - 2T_\mathcal{X} F_\mathcal{X} = 0$, i.e., $T_\mathcal{X} F_\mathcal{X} = \frac{1}{2}\big[\kappa_\mathcal{X}^{(T)}\big]^2$.

From the above, it is clear that the minimum $T_{\mathcal{X}} F_{\mathcal{X}}$ will be achieved if $\mathcal{X} = \mathcal{H}_N^{(T_0)}$. However, it is easily shown[6] that the product $T_{\mathcal{H}_N^{(T_0)}} F_{\mathcal{H}_N^{(T_0)}}$ is independent of T_0 (even though $T_{\mathcal{H}_N^{(T_0)}}$ and $F_{\mathcal{H}_N^{(T_0)}}$ individually depend on T_0), and hence the minimum $T_{\mathcal{X}} F_{\mathcal{X}}$ will be achieved for $\mathcal{X} = \mathcal{H}_N^{(T)}$ where T is arbitrary.

Appendix 3.B: Proof of Theorem 3.2

In what follows, let $\mathcal{T} \triangleq \mathcal{T}[-T/2, T/2]$ and $\mathcal{F} \triangleq \mathcal{F}[-F/2, F/2]$ for brevity. We first recall that the prolate spheroidal wave functions $p_k^{(T,F)}(t)$ are the eigenfunctions of the linear operator $\mathbf{K}^{(T,F)} = \mathbf{P}_{\mathcal{F}} \mathbf{P}_{\mathcal{T}}$ obtained by composing the orthogonal projection operator $\mathbf{P}_{\mathcal{T}}$ (the time-limitation operator on the time interval $[-T/2, T/2]$) with the orthogonal projection operator $\mathbf{P}_{\mathcal{F}}$ (the band-limitation operator, or idealized lowpass filter, on the frequency band $[-F/2, F/2]$). The $p_k^{(T,F)}(t)$ are arranged such that the corresponding eigenvalues $\lambda_k^{(T,F)}$ are monotonically decreasing, $1 > \lambda_1^{(T,F)} > \lambda_2^{(T,F)} > \ldots > 0$. Since the $p_k^{(T,F)}(t)$ are themselves band-limited to $[-F/2, F/2]$, there is $\left(\mathbf{P}_{\mathcal{F}} p_k^{(T,F)}\right)(t) = p_k^{(T,F)}(t)$, and it is easily shown that the $\lambda_k^{(T,F)}$ and $p_k^{(T,F)}(t)$ are also the eigenvalues and eigenfunctions, respectively, of the self-adjoint operator[7]

$$\left[\mathbf{P}_{\mathcal{T}}\right]_{\mathcal{F}} \triangleq \mathbf{P}_{\mathcal{F}} \mathbf{P}_{\mathcal{T}} \mathbf{P}_{\mathcal{F}},$$

which is the $\mathcal{F}$-restricted time-limitation operator $\mathbf{P}_{\mathcal{T}}$ (cf. Section 2.7.2).

The temporal concentration $\alpha_x^{(T)} = E_x^{(T)} / E_x$ of a signal $x(t)$ is a normalized version of the interval energy

$$E_x^{(T)} = \int_{-T/2}^{T/2} d_x(t)\, dt = \langle \mathbf{P}_{\mathcal{T}}\, x,\, x \rangle.$$

Similarly, the temporal concentration of a signal space $\mathcal{X}$ is a normalized version

---

[6]Indeed, it follows from $T^2_{\mathcal{H}_N^{(T)}} = \frac{1}{N} \sum_{k=1}^{N} T^2_{h_k^{(T)}}$ and $h_k^{(T)}(t) = \frac{1}{\sqrt{T}} h_k\!\left(\frac{t}{T}\right)$ that

$$T_{\mathcal{H}_N^{(T)}} = T\sqrt{m} \qquad \text{where} \quad m = \frac{1}{N} \sum_{k=1}^{N} m_{h_k}^{(2)},$$

i.e., $T_{\mathcal{H}_N^{(T)}}$ is proportional to T. An analogous argument shows that $F_{\mathcal{H}_N^{(T)}}$ is inversely proportional to T. Hence, in the product $T_{\mathcal{H}_N^{(T)}} F_{\mathcal{H}_N^{(T)}}$ the parameter T cancels.

[7]Note that, even though the eigenvalues and eigenfunctions of the operators $\mathbf{P}_{\mathcal{F}} \mathbf{P}_{\mathcal{T}} \mathbf{P}_{\mathcal{F}}$ and $\mathbf{P}_{\mathcal{F}} \mathbf{P}_{\mathcal{T}}$ are identical, the operators themselves are quite different (e.g., $\mathbf{P}_{\mathcal{F}} \mathbf{P}_{\mathcal{T}} \mathbf{P}_{\mathcal{F}}$ is a self-adjoint operator whereas $\mathbf{P}_{\mathcal{F}} \mathbf{P}_{\mathcal{T}}$ is not even normal [Naylor and Sell, 1982]).

of the interval energy $E_{\mathcal{X}}^{(T)}$,

$$\alpha_{\mathcal{X}}^{(T)} = \frac{E_{\mathcal{X}}^{(T)}}{N_{\mathcal{X}}},$$

where

$$E_{\mathcal{X}}^{(T)} = \int_{-T/2}^{T/2} d_{\mathcal{X}}(t)\, dt = \sum_{k=1}^{N_{\mathcal{X}}} E_{x_k}^{(T)} = \sum_{k=1}^{N_{\mathcal{X}}} \langle \mathbf{P}_{\mathcal{T}} x_k, x_k \rangle . \tag{3.B.1}$$

We now look for the N-dimensional space $\mathcal{X}$ maximizing the temporal concentration $\alpha_{\mathcal{X}}^{(T)}$ under the side constraint that $\mathcal{X}$ is band-limited to $[-F/2, F/2]$, i.e., $\mathcal{X} \subseteq \mathcal{F}$. Since the dimension $N_{\mathcal{X}} = N$ is fixed, an equivalent problem is the maximization of the interval energy $E_{\mathcal{X}}^{(T)}$, i.e., the optimum space is

$$\mathcal{X}_{\mathrm{opt}}(N) = \arg \max_{\substack{N_{\mathcal{X}}=N \\ \mathcal{X} \subseteq \mathcal{F}}} E_{\mathcal{X}}^{(T)} .$$

With (3.B.1), the results of Theorem 2.2 (see Section 2.7.2) are directly applicable to our maximization problem, and it follows that $\mathcal{X}_{\mathrm{opt}}(N)$ is spanned by the first N eigenfunctions of $\left[\mathbf{P}_{\mathcal{T}}\right]_{\mathcal{F}}$. But we have shown above that the eigenfunctions of $\left[\mathbf{P}_{\mathcal{T}}\right]_{\mathcal{F}}$ are just the $p_k^{(T,F)}(t)$. Hence, the optimum space is the prolate spheroidal space of dimension N,

$$\mathcal{X}_{\mathrm{opt}}(N) = \mathrm{span}\{p_k^{(T,F)}(t)\}_{k=1}^{N} = \mathcal{P}_N^{(T,F)} ,$$

as claimed in Theorem 3.2. Furthermore, the maximum interval energy is (see Theorem 2.2)

$$E_{\mathcal{X}}^{(T)}\bigg|_{\mathrm{max}} = \sum_{k=1}^{N} \lambda_k^{(T,F)} ,$$

i.e., the sum of the N first (largest) eigenvalues of $\left[\mathbf{P}_{\mathcal{T}}\right]_{\mathcal{F}}$ or, equivalently, of $\mathbf{K}^{(T,F)}$. The maximum temporal concentration is finally obtained as

$$\alpha_{\mathcal{X}}^{(T)}\bigg|_{\mathrm{max}} = \alpha_{\mathcal{X}}^{(T)}\bigg|_{\mathcal{X}=\mathcal{P}_N^{(T,F)}} = \frac{1}{N} E_{\mathcal{X}}^{(T)}\bigg|_{\mathrm{max}} = \frac{1}{N} \sum_{k=1}^{N} \lambda_k^{(T,F)} ,$$

which is the lower bound (3.15) to be proved.

4   TIME-FREQUENCY SYNTHESIS OF LINEAR SIGNAL SPACES

So far, we have considered the *TF analysis* of linear signal spaces by means of the WD of a linear signal space. In this chapter, we present a framework and a method for the optimum *TF synthesis* of linear signal spaces using the WD. The synthesis problem is formulated as finding the space with optimum localization in a given region of the TF plane. In other words, all signals of the optimum space are effectively located inside the given TF region, in such a manner that the TF region is "filled energetically" by the space while only a minimum amount of space energy leaks to the outside of the TF region. Mathematically, this synthesis problem can be formulated as a minimization of the space's TF localization error as introduced in the previous chapter.

A variation of our space synthesis method includes a *subspace side constraint* whereby the space to be synthesized is constrained to be a subspace of a given signal (sub-)space. For example, such a subspace-constrained synthesis is appropriate if the space to be synthesized is required to be strictly band-limited to a given frequency band.

The optimum TF synthesis discussed in this chapter will be applied to the design of *TF projection filters* in Chapter 5. These filters pass (suppress) all signals located inside (outside) a given TF pass region. A related application, also to be discussed in Chapter 5, is a *TF signal expansion* allowing the parsimonious representation of signals located within a given TF support region.

This chapter is organized as follows. Section 4.1 formulates the TF synthesis problem as a minimization of the TF localization error and shows that the optimum space is an "eigenspace" of the given TF region. Some fundamental properties of such eigenspaces are then discussed in Section 4.2. In Section 4.3, the TF synthesis problem and its solution are extended to include a subspace constraint. Discrete-time implementations of the synthesis methods are proposed in Section 4.4, and simulation results are presented in Section 4.5. Finally, Section 4.6 shows that the two space families that have been found in Chapter 3 to feature optimum TF concentration are also optimum in the context of TF synthesis.

4.1 Optimum Time-Frequency Synthesis

The TF synthesis problem considered in this section is verbally stated as follows: Construct a linear signal space ("TF space") $\mathcal{X}_R$ which "comprises all signals located in a given TF region R." A slightly more precise specification is that the WD of the TF space $\mathcal{X}_R$ be optimally localized in R, in the sense that it covers R energetically but has little or no energy outside R.

4.1.1 Optimality Criteria

Based on the above specification, it is natural to define the optimum space as the space minimizing the TF localization error $\epsilon(\mathcal{X}, R)$ defined in (3.19). According to this *criterion of minimum TF localization error* (MLE), the optimum TF space is [Hlawatsch and Kozek, 1994]

$$\mathcal{X}_{R,\mathrm{MLE}} \;\overset{\triangle}{=}\; \arg\min_{\mathcal{X}} \; \epsilon(\mathcal{X}, R) \;=\; \arg\min_{\mathcal{X}} \; \|I_R - W_{\mathcal{X}}\| \,,$$

where $I_R(t, f)$ is the indicator function of the TF region R and $\|I_R - W_{\mathcal{X}}\|^2 = \int_t \int_f \left[I_R(t, f) - W_{\mathcal{X}}(t, f) \right]^2 dt\, df$. Thus, the optimum TF space is defined as the space whose WD is closest to the TF region's indicator function.

An alternative, intuitively appealing optimality criterion is the *criterion of maximum regional TF concentration* (MRC), where the optimum TF space is defined as the space with maximum regional TF concentration $\rho(\mathcal{X}, R)$ (see (3.16)). However, the regional TF concentration has been seen in Section 3.5.2 to be insensitive to energy gaps inside R. Indeed, the concentration inequality (3.18) shows that good regional TF concentration is favored by a small space dimension. It will be seen later that the MRC criterion always results in a one-dimensional subspace, which is generally not a desired result (according to the localization inequality (3.21), the space dimension should be approximately equal to the TF region's area). Therefore, the MRC criterion is meaningful only if the space's dimension is fixed beforehand, $N_\mathcal{X} = N$, and we thus define the MRC space *of given dimension N* as

$$\mathcal{X}_{R,\mathrm{MRC}}(N) \;\overset{\triangle}{=}\; \arg\max_{N_\mathcal{X}=N} \rho(\mathcal{X}, R) \;=\; \arg\max_{N_\mathcal{X}=N} \langle W_\mathcal{X}, I_R \rangle \,.$$

Note that the MRC criterion leaves the dimension $N_\mathcal{X} = N$ to be chosen, whereas the MLE criterion yields an optimum dimension.

From (3.20), it follows that *for prescribed (i.e., fixed) dimension of the space, $N_\mathcal{X} = N$, minimization of the TF localization error is equivalent to maximization of the regional TF concentration,*

$$\mathcal{X}_{R,\mathrm{MLE}}(N) = \mathcal{X}_{R,\mathrm{MRC}}(N) \,.$$

Here, $\mathcal{X}_{R,\mathrm{MLE}}(N)$ denotes the space with minimum TF localization error subject to a given space dimension N. Note, however, that we normally use the MLE criterion without fixing the dimension beforehand.

4.1.2 Eigenspaces of a Time-Frequency Region

The calculation of the MLE and MRC spaces is based on the results derived in Section 2.7.1. Indeed, we can associate the indicator function $I_R(t, f)$ of the TF region R with a self-adjoint linear operator $\mathbf{H}_R$ whose kernel $H_R(t_1, t_2)$ is defined by the Weyl symbol relation (cf. (2.58), (2.59))

$$I_R(t, f) \;=\; \int_\tau H_R\left(t + \frac{\tau}{2}, t - \frac{\tau}{2}\right) e^{-j2\pi f\tau}\, d\tau \,, \tag{4.1}$$

$$H_R(t_1, t_2) \;=\; \int_f I_R\left(\frac{t_1 + t_2}{2}, f\right) e^{j2\pi(t_1 - t_2)f}\, df \,. \tag{4.2}$$

The mapping $H_R(t_1, t_2) \longleftrightarrow I_R(t, f)$ defined above is linear and unitary. Since $I_R(t, f)$ is real-valued and square-integrable (for bounded TF region R), the function $H_R(t_1, t_2)$ is Hermitian, i.e., $H_R^*(t_2, t_1) = H_R(t_1, t_2)$, and square-integrable. Thus, the operator $\mathbf{H}_R$ is self-adjoint and compact [Naylor and Sell, 1982]. This guarantees the existence of the eigenexpansion

$$H_R(t_1, t_2) \;=\; \sum_{k=1}^{\infty} \lambda_k\, u_k(t_1)\, u_k^*(t_2) \tag{4.3}$$

with a discrete spectrum of real-valued eigenvalues λ_k and orthonormal eigenfunctions $u_k(t)$ defined by the eigenequation

$$\int_{t_2} H_R(t_1, t_2)\, u_k(t_2)\, dt_2 \;=\; \lambda_k\, u_k(t_1) \,.$$

Inserting (4.3) into (4.1) gives the expansion

$$I_R(t, f) = \sum_{k=1}^{\infty} \lambda_k\, W_{u_k}(t, f)\,, \tag{4.4}$$

which is a special case of (2.64). The real-valued coefficients λ_k and the orthonormal signals $u_k(t)$ are completely determined by the TF region R; they will hence be called *eigenvalues and eigensignals of the TF region R*, respectively. Using these eigenvalues and eigensignals, we next define the eigenspaces of a TF region.

Definition 4.1 *The N-dimensional eigenspace $\mathcal{U}_R^{(N)}$ of a TF region R is defined as the space spanned by the N dominant eigensignals of R, i.e., the N eigensignals $u_k(t)$ with largest eigenvalues λ_k.*

In what follows, we assume that the eigenvalues are arranged in non-increasing order, $\lambda_1 \geq \lambda_2 \geq \dots$. Then, $\mathcal{U}_R^{(N)}$ is spanned by the *first N* eigensignals,

$$\mathcal{U}_R^{(N)} = \operatorname{span}\{u_k(t)\}_{k=1}^{N}\,.$$

The eigenspaces of a TF region provide the solution to both the MLE and MRC optimization problems, as is stated by the following theorem [Hlawatsch et al., 1990, Hlawatsch and Kozek, 1994].

Theorem 4.1 *The N-dimensional signal space minimizing the TF localization error $\epsilon(\mathcal{X}, R)$ and maximizing the regional TF concentration $\rho(\mathcal{X}, R)$ for a given TF region R is the eigenspace of R with dimension N,*

$$\mathcal{X}_{R,\mathrm{MLE}}(N) = \mathcal{X}_{R,\mathrm{MRC}}(N) = \mathcal{U}_R^{(N)}\,.$$

Assuming that the eigenvalues λ_k are arranged in non-increasing order, the resulting TF localization error (the minimum TF localization error for the given dimension N) is related to the eigenvalues of R as

$$\epsilon^2(\mathcal{X}, R)\Big|_{\min,N} = \epsilon^2(\mathcal{X}, R)\Big|_{\mathcal{X}=\mathcal{U}_R^{(N)}} = \sum_{k=1}^{N}(1 - \lambda_k)^2 + \sum_{k=N+1}^{\infty} \lambda_k^2\,, \tag{4.5}$$

and the resulting regional TF concentration (the maximum regional TF concentration for the given dimension N) is the arithmetic mean of the N first (largest) eigenvalues of R,

$$\rho(\mathcal{X}, R)\Big|_{\max,N} = \rho(\mathcal{X}, R)\Big|_{\mathcal{X}=\mathcal{U}_R^{(N)}} = \frac{1}{N}\sum_{k=1}^{N}\lambda_k\,. \tag{4.6}$$

Proof. Since $N_{\mathcal{X}} = N$ is fixed, maximization of $\rho(\mathcal{X}, R) = E_{\mathcal{X}}^{(R)}/N_{\mathcal{X}}$ is equivalent to maximization of the spaces' regional energy $E_{\mathcal{X}}^{(R)} = \langle W_{\mathcal{X}}, I_R \rangle$. Hence, Theorem 2.1 (see Section 2.7.1) is directly applicable if we set $G(t, f) = I_R(t, f)$ and $\mathbf{H} = \mathbf{H}_R$. This shows that the N-dimensional space maximizing $E_{\mathcal{X}}^{(R)}$ is spanned by the N first (i.e., largest) eigenfunctions of $\mathbf{H}_R$. These are the N first (i.e., largest) eigensignals of the TF region R, and thus the optimum space is $\mathcal{U}_R^{(N)}$. According to Theorem 2.1, the regional energy achieved equals the sum of the N first (i.e., largest) eigenvalues, $E_{\mathcal{X}}^{(R)} = \sum_{k=1}^{N} \lambda_k$, which yields (4.6). Finally, (4.5) is easily derived from (4.6) with the use of (3.20) and the relation $A_R = \sum_{k=1}^{\infty} \lambda_k^2$ that will be proved in Section 4.2.2 (see (4.9)). $\square$

In Theorem 4.1, we have prescribed the dimension N also for the MLE optimization problem. However, since (4.5) gives the residual TF localization error for any dimension N, the optimum value of N can be determined by minimizing (4.5) with respect to N. It is seen by inspection that the optimum N equals the number of all eigenvalues larger than $1/2$. This yields the final result for the MLE subspace:

Corollary 4.1 *The signal space minimizing the TF localization error $\epsilon(\mathcal{X}, R)$ for a given TF region R is the eigenspace of R with dimension N_R,*

$$\mathcal{X}_{R,\mathrm{MLE}} = \mathcal{U}_R^{(N_R)},$$

where N_R denotes the number of eigenvalues λ_k larger than $1/2$. The residual (absolutely minimum) TF localization error is

$$\epsilon^2(\mathcal{X}, R)\Big|_{\min} = \epsilon^2(\mathcal{X}, R)\Big|_{\mathcal{X}=\mathcal{U}_R^{(N_R)}} = \sum_{k=1}^{N_R}(1 - \lambda_k)^2 + \sum_{k=N_R+1}^{\infty} \lambda_k^2 . \qquad (4.7)$$

In contrast to the MLE criterion, the MRC criterion cannot be used for determining an optimum dimension N in a meaningful manner. However, based on the expression (4.6) of $\rho(\mathcal{X}, R)\big|_{\max,N}$ in terms of the eigenvalues λ_k, the dimension N can easily be adjusted such that a prescribed regional TF concentration is either approximately achieved or exceeded: we simply have to select the N for which the arithmetic mean of the first N eigenvalues is, respectively, closest to or larger than the regional TF concentration specified.

From (4.6) and the monotonicity $\lambda_k \geq \lambda_{k+1}$ of the eigenvalues, it is clear that the regional TF concentration achieved will decrease with increasing dimension N. The absolutely maximum regional TF concentration is obtained for $N = 1$. Of course, a one-dimensional signal space is generally incapable of filling the region R energetically, and thus does not solve our problem. On the other hand, selecting N too large results in a space which "spills over" the TF region R, causing a small regional concentration value. We shall usually adopt the optimum dimension $N = N_R$; the corresponding eigenspace $\mathcal{U}_R^{(N_R)}$ will then simply be called "the eigenspace of R" and will be denoted as $\mathcal{U}_R$. Similarly, the residual (absolutely minimum) TF localization error in (4.7) will be called "the TF localization error associated with R" and will be denoted as ϵ_R:

$$\mathcal{U}_R \stackrel{\triangle}{=} \mathcal{U}_R^{(N_R)}, \qquad \epsilon_R \stackrel{\triangle}{=} \epsilon(\mathcal{U}_R, R) \,.$$

We note that $\mathcal{U}_R$, ϵ_R, and N_R are completely determined by the TF region R. With the definition of the "idealized eigenvalues"

$$\tilde{\lambda}_k \stackrel{\triangle}{=} \begin{cases} 1, & 1 \leq k \leq N_R \\ 0, & N_R + 1 \leq k < \infty, \end{cases}$$

the squared TF localization error ϵ_R in (4.7) can be compactly written as the deviation between the actual eigenvalues and the idealized eigenvalues,

$$\epsilon_R^2 = \sum_{k=1}^{N_R}(1 - \lambda_k)^2 + \sum_{k=N_R+1}^{\infty} \lambda_k^2 = \sum_{k=1}^{\infty}(\tilde{\lambda}_k - \lambda_k)^2 \,.$$

Note that ϵ_R would be zero if the eigenvalues λ_k assumed only the values 0 or 1 (i.e., if $\lambda_k = \tilde{\lambda}_k$); however, this idealized eigenvalue distribution is never obtained for a bounded TF region R.

The MLE problem $\mathcal{X}_{R,\mathrm{MLE}} = \arg\min_{\mathcal{X}} \|I_R - W_{\mathcal{X}}\|$ is the optimum approximation of a "non-valid" TF function $I_R(t, f)$ ("non-valid" in the sense that $I_R(t, f)$ is not a valid WD of a space) by a valid WD of a space. This is analogous to the *signal synthesis problem* $x_{\mathrm{opt}}(t) = \arg\min_x \|\widetilde{W} - W_x\|$ where a "non-valid" TF function $\widetilde{W}(t, f)$ ("non-valid" in the sense that $\widetilde{W}(t, f)$ is not a valid WD of a signal) is approximated by a valid WD of a signal [Boudreaux-Bartels and Parks, 1986, Saleh and Subotic, 1985, Hlawatsch and Krattenthaler, 1992, Hlawatsch and Krattenthaler, 1997, Boudreaux-Bartels, 1997]. In fact, the signal synthesis problem is essentially the space synthesis problem with $N = 1$. In a similar manner, the MRC problem $\mathcal{X}_{R,\mathrm{MRC}} = \arg\max_{\mathcal{X}} \langle W_{\mathcal{X}}, I_R \rangle$ can be viewed as an extension of the problem of maximizing the regional TF concentration of a signal. The latter problem, which is essentially equivalent to the signal synthesis problem, has been studied in [Flandrin, 1988].

4.2   Properties of Eigenspaces

We have shown above that the optimum TF space defined by the MLE or MRC criterion is an eigenspace $\mathcal{U}_R^{(N)}$ of the TF region R. This space is spanned by a set of dominant eigensignals $u_k(t)$, and the dimension N chosen or obtained depends on the distribution of the eigenvalues λ_k. We shall now study some properties of the eigenvalues, eigensignals, and eigenspaces of a TF region.

4.2.1 Concentration Bounds

We first recall that the regional TF concentration $\rho(x, R)$ of a signal $x(t)$ in a TF region R has been defined in Section 3.5.3 as

$$\rho(x, R) \;=\; \frac{E_x^{(R)}}{E_x} \;=\; \frac{\iint_{(t,f)\in R} W_x(t, f)\, dt\, df}{\int_t \int_f W_x(t, f)\, dt\, df} \;=\; \frac{1}{E_x}\, \langle W_x, I_R \rangle \,.$$

The regional TF concentration of a space $\mathcal{X}$ can be shown to be the mean of the regional TF concentrations of the space's orthonormal basis signals $x_k(t)$,

$$\rho(\mathcal{X}, R) \;=\; \frac{1}{N_{\mathcal{X}}} \sum_{k=1}^{N_{\mathcal{X}}} \rho(x_k, R) \,.$$

Furthermore, according to (2.63) the inner product of the cross-WD of the eigensignals and the indicator function satisfies

$$\langle W_{u_k, u_l}, I_R \rangle \;=\; \iint_{(t,f)\in R} W_{u_k, u_l}(t, f)\, dt\, df \;=\; \lambda_k \delta_{kl} \,. \tag{4.8}$$

In particular, the regional TF concentrations of the eigensignals $u_k(t)$ equal the eigenvalues λ_k,

$$\rho(u_k, R) \;=\; \langle W_{u_k}, I_R \rangle \;=\; \iint_{(t,f)\in R} W_{u_k}(t, f)\, dt\, df \;=\; \lambda_k \,.$$

The next theorem establishes bounds on the regional TF concentration of the N-dimensional eigenspace $\mathcal{U}_R^{(N)}$ and any of its elements.

Theorem 4.2 *The regional TF concentration of the N-dimensional eigenspace $\mathcal{U}_R^{(N)}$ is bounded as*

$$\lambda_N \;\leq\; \rho(\mathcal{U}_R^{(N)}, R) \;\leq\; \lambda_1 \,.$$

The regional TF concentration of any element $x(t)$ of $\mathcal{U}_R^{(N)}$ is bounded as

$$\lambda_N \;\leq\; \rho(x, R) \;\leq\; \lambda_1 \qquad \text{for all } x(t) \in \mathcal{U}_R^{(N)} \,.$$

The latter lower and upper bounds are tight since they are attained by $u_N(t)$ and $u_1(t)$, respectively, both of which are elements of $\mathcal{U}_R^{(N)}$.

Proof. The bounds on $\rho(\mathcal{U}_R^{(N)}, R)$ follow from (4.6), stating that $\rho(\mathcal{U}_R^{(N)}, R)$ is the mean of the N first (i.e., largest) eigenvalues. Due to $\lambda_{k+1} \leq \lambda_k$, this mean is bounded from below by λ_N and from above by λ_1.

In order to show the bounds on $\rho(x, R)$, we note that any $x(t) \in \mathcal{U}_R^{(N)}$ can be expanded as $x(t) = \sum_{k=1}^{N} \alpha_k\, u_k(t)$, from which it follows that the WD of $x(t)$ can be written as

$$W_x(t, f) = \sum_{k=1}^{N} \sum_{l=1}^{N} \alpha_k \alpha_l^* \, W_{u_k, u_l}(t, f) \, .$$

The regional energy then becomes

$$E_x^{(R)} = \langle W_x, I_R \rangle = \sum_{k=1}^{N} \sum_{l=1}^{N} \alpha_k \alpha_l^* \langle W_{u_k, u_l}, I_R \rangle = \sum_{k=1}^{N} \sum_{l=1}^{N} \alpha_k \alpha_l^* \, \lambda_k \, \delta_{kl}$$

$$= \sum_{k=1}^{N} \lambda_k \, |\alpha_k|^2 \, ,$$

where (4.8) has been used. With the energy of $x(t)$ being $E_x = \sum_{k=1}^{N} |\alpha_k|^2$, the regional TF concentration of $x(t)$ is

$$\rho(x, R) = \frac{E_x^{(R)}}{E_x} = \frac{\sum_{k=1}^{N} \lambda_k \, |\alpha_k|^2}{\sum_{k=1}^{N} |\alpha_k|^2} \geq \lambda_N \, \frac{\sum_{k=1}^{N} |\alpha_k|^2}{\sum_{k=1}^{N} |\alpha_k|^2} = \lambda_N \, .$$

The upper bound follows by an analogous argument. $\qquad\square$

The eigenspace $\mathcal{U}_R$ (where $N = N_R$) is of particular interest. Here, $\lambda_N = \lambda_{N_R} > 1/2$, so that

$$\rho(\mathcal{U}_R, R) > 1/2 \qquad \text{and} \qquad \rho(x, R) > 1/2 \quad \text{for all} \ \ x(t) \in \mathcal{U}_R \, .$$

4.2.2 Asymptotic Properties

The eigenvalue distribution obtained for a TF region R determines the optimum dimension N_R and the dimension necessary for achieving a desired regional TF concentration. The eigenvalues also determine the TF localization error ϵ_R. An interesting property of the eigenvalues is stated in the next theorem.

Theorem 4.3 *The sum of all eigenvalues and the sum of all squared eigenvalues are both equal to the area A_R,*

$$\sum_{k=1}^{\infty} \lambda_k = \sum_{k=1}^{\infty} \lambda_k^2 = A_R \, . \tag{4.9}$$

Proof. Using (3.17), (4.4), and Moyal's formula (1.10), we have

$$A_R = \int_t \int_f I_R(t, f) \, dt \, df = \sum_{k=1}^{\infty} \lambda_k \int_t \int_f W_{u_k}(t, f) \, dt \, df$$

$$= \sum_{k=1}^{\infty} \lambda_k \, \|u_k\|^2 = \sum_{k=1}^{\infty} \lambda_k \, ;$$

$$A_R = \int_t \int_f [I_R(t, f)]^2 \, dt \, df = \sum_{k=1}^{\infty} \sum_{l=1}^{\infty} \lambda_k \, \lambda_l \, \langle W_{u_k}, W_{u_l} \rangle$$

$$= \sum_{k=1}^{\infty} \sum_{l=1}^{\infty} \lambda_k \, \lambda_l \, |\langle u_k, u_l \rangle|^2 = \sum_{k=1}^{\infty} \lambda_k^2 \, . \qquad \square$$

In [Ramanathan and Topiwala, 1993, Heil et al., 1994], the asymptotic behavior of the eigenvalues and eigensignals is studied for a bounded TF region R. It is shown that $|\lambda_k|$ decays as $\mathcal{O}(k^{-3/4})$ but not as $\mathcal{O}(k^{-(1+\epsilon)})$ for any $\epsilon > 0$, and that $\sum_k |\lambda_k| = \infty$. (Note, however, that $\sum_k \lambda_k = \sum_k \lambda_k^2 = A_R < \infty$.) Furthermore, the eigensignals corresponding to nonzero eigenvalues are shown to have faster than exponential decay in both time domain and frequency domain: for any $a > 0$, there exist positive constants c_a and C_a such that $|u_k(t)| \leq c_a \, e^{-a|t|}$ for all t and $|U_k(f)| \leq C_a \, e^{-a|f|}$ for all f, where $U_k(f)$ is the Fourier transform of $u_k(t)$. Moreover, both $u_k(t)$ and $U_k(f)$ are analytic functions, i.e., all derivatives exist and are in $\mathcal{L}_2(\mathbb{R})$. These results show that the error introduced by a temporal truncation or band-limitation of the eigensignals $u_k(t)$ will be negligibly small if the length of the time-gating window, or the bandwidth of the filter, is chosen sufficiently large.

4.2.3   Covariance Property

In Section 2.3, we have discussed a class of unitary, linear signal or space transformations $\mathbf{H}$ corresponding to area-preserving affine TF coordinate transforms

$$(t, f) \longrightarrow (\alpha t + \beta f - \tau, \, \gamma t + \delta f - \nu) \qquad \text{with} \quad D = \alpha\delta - \gamma\beta = 1 \, ,$$

where the parameters $\alpha, \beta, \gamma, \delta, \tau$, and ν depend on the signal/space transformation $\mathbf{H}$, and the area preservation implies a transform determinant $D = 1$ [Janssen, 1982]. The WD of a transformed signal or space is related to the WD of the original signal or space according to the following covariance property:

$$W_{\mathbf{H}x}(t, f) = W_x(\alpha t + \beta f - \tau, \, \gamma t + \delta f - \nu) \, , \qquad (4.10)$$

$$W_{\mathbf{H}\chi}(t, f) = W_\chi(\alpha t + \beta f - \tau, \, \gamma t + \delta f - \nu) \, . \qquad (4.11)$$

We recall that some important special cases are the following:

$$(\mathbf{H}x)(t) = x(t-\tau)\,e^{j2\pi\nu t} \quad \Rightarrow \quad (t,f) \longrightarrow (t-\tau, f-\nu)$$
$$(\mathbf{H}x)(t) = \sqrt{|a|}\,x(at) \quad \Rightarrow \quad (t,f) \longrightarrow (at, f/a)$$
$$(\mathbf{H}x)(t) = x(t) * \sqrt{|c|}\,e^{j\pi ct^2} \quad \Rightarrow \quad (t,f) \longrightarrow (t-f/c, f)$$
$$(\mathbf{H}x)(t) = x(t)\,e^{j\pi ct^2} \quad \Rightarrow \quad (t,f) \longrightarrow (t, f-ct)$$
$$(\mathbf{H}x)(t) = \sqrt{|c|}\,X(ct) \quad \Rightarrow \quad (t,f) \longrightarrow (-f/c, ct)\,.$$

Since a general affine coordinate transform can be represented by composing some of the specific transforms listed above [Papoulis, 1984b], the operator $\mathbf{H}$ in (4.10), (4.11) can be constructed for any set of parameters $\alpha, \beta, \gamma, \delta, \tau, \nu$ satisfying $D = 1$ [Janssen, 1982].

The above covariance property of the WD of signals or spaces holds "in an inverse sense" also for the synthesis of spaces. If we subject the TF plane to an area-preserving affine coordinate transform $(t, f) \to (\alpha t + \beta f - \tau, \gamma t + \delta f - \nu)$, the TF region R maps into a new region $\tilde{R}$ with indicator function

$$I_{\tilde{R}}(t,f) \;=\; I_R(\alpha t + \beta f - \tau, \gamma t + \delta f - \nu)\,.$$

Due to area preservation, the region's area is unchanged,

$$A_{\tilde{R}} \;=\; A_R\,.$$

The next theorem (cf. [Folland, 1989], p. 83 f.) shows how the eigenspaces of the original region R and the new region $\tilde{R}$ are related.

Theorem 4.4 *Let $\tilde{R}$ be the TF region obtained by subjecting a given TF region R to an affine TF coordinate transform $(t, f) \longrightarrow (\alpha t + \beta f - \tau, \gamma t + \delta f - \nu)$ with $\alpha\delta - \gamma\beta = 1$. Then, the eigenvalues of $\tilde{R}$ equal those of R, whereas the eigensignals of $\tilde{R}$ are derived from those of R by the unitary, linear signal transformation $\mathbf{H}$ corresponding to the given TF coordinate transform:*

$$\tilde{\lambda}_k \;=\; \lambda_k\,, \qquad \tilde{u}_k(t) \;=\; (\mathbf{H}u_k)(t)\,.$$

The eigenspace $\mathcal{U}_R^{(N)} = \mathrm{span}\{u_k(t)\}_{k=1}^N$ of the original TF region R and the eigenspace $\mathcal{U}_{\tilde{R}}^{(N)} = \mathrm{span}\{\tilde{u}_k(t)\}_{k=1}^N$ of the transformed TF region $\tilde{R}$ are related via the transformation $\mathbf{H}$ (in the sense of Section 2.3),

$$\mathcal{U}_{\tilde{R}}^{(N)} \;=\; \mathbf{H}\,\mathcal{U}_R^{(N)}\,, \tag{4.12}$$

and the WDs of these eigenspaces are related by the given coordinate transform,

$$W_{\mathcal{U}_{\tilde{R}}^{(N)}}(t,f) \;=\; W_{\mathcal{U}_R^{(N)}}(\alpha t + \beta f - \tau, \gamma t + \delta f - \nu)\,. \tag{4.13}$$

Proof. Using (4.4) and (4.10), the indicator function of the transformed TF region $\tilde{R}$ becomes

$$
\begin{aligned}
I_{\tilde{R}}(t,f) &= I_R(\alpha t + \beta f - \tau, \gamma t + \delta f - \nu) \\
&= \sum_{k=1}^{\infty} \lambda_k \, W_{u_k}(\alpha t + \beta f - \tau, \gamma t + \delta f - \nu) = \sum_{k=1}^{\infty} \lambda_k \, W_{\mathbf{H}u_k}(t,f) \,.
\end{aligned}
$$

Since $\mathbf{H}$ is unitary, the transformed eigensignals $(\mathbf{H}u_k)(t)$ are again orthonormal, so that λ_k and $(\mathbf{H}u_k)(t)$ are recognized as the eigenvalues and eigensignals, respectively, of $\tilde{R}$, i.e., $\tilde{\lambda}_k = \lambda_k$ and $\tilde{u}_k(t) = (\mathbf{H}u_k)(t)$. Relation (4.12) follows readily from the fact that $\mathcal{U}_{\tilde{R}}^{(N)}$ is spanned by the orthonormal basis $\tilde{u}_k(t) = (\mathbf{H}u_k)(t)$. With (2.7), relation (4.13) is finally shown as follows:

$$
\begin{aligned}
W_{\mathcal{U}_{\tilde{R}}^{(N)}}(t,f) &= \sum_{k=1}^{N} W_{\mathbf{H}u_k}(t,f) = \sum_{k=1}^{N} W_{u_k}(\alpha t + \beta f - \tau, \gamma t + \delta f - \nu) \\
&= W_{\mathcal{U}_R^{(N)}}(\alpha t + \beta f - \tau, \gamma t + \delta f - \nu) \,. \qquad \square
\end{aligned}
$$

From Theorem 4.4, we can draw an important conclusion: Since the eigenvalues are not affected by the coordinate transform, $\tilde{\lambda}_k = \lambda_k$, and since the optimum dimension N_R and the minimum TF localization error ϵ_R are determined by the eigenvalues, the coordinate transform does not change the optimum dimension or the minimum TF localization error. Thus, we have

$$
N_{\tilde{R}} = N_R \,, \qquad \epsilon_{\tilde{R}} = \epsilon_R \,.
$$

4.2.4 Complement Property

A further interesting property of eigenspaces is stated in the next theorem. This property will be seen to be useful for the design of perfect-reconstruction TF filter banks discussed in Section 5.3.

Theorem 4.5 *Let R be a TF region for which $1/2$ is not an eigenvalue. Then, the eigenspace $\mathcal{U}_R$ and the eigenspace $\mathcal{U}_{\bar{R}}$ of the complementary TF region $\bar{R} = \mathbb{R}^2 \setminus R$ are orthogonal complement spaces,*

$$
\mathcal{U}_{\bar{R}} = \overline{\mathcal{U}_R} \,,
$$

which means that $\mathcal{U}_R$ and $\mathcal{U}_{\bar{R}}$ are orthogonal and their (direct) sum is $\mathcal{L}_2(\mathbb{R}^2)$,

$$
\mathcal{U}_R \perp \mathcal{U}_{\bar{R}} \qquad and \qquad \mathcal{U}_R + \mathcal{U}_{\bar{R}} = \mathcal{L}_2(\mathbb{R}^2) \,.
$$

Proof. The indicator functions of R and $\bar{R}$ satisfy $I_R(t,f) + I_{\bar{R}}(t,f) \equiv 1$ or

$$I_{\bar{R}}(t,f) = 1 - I_R(t,f) \,.$$

Applying the inverse Weyl transform (4.2) yields

$$
H_{\bar{R}}(t_1,t_2) = \delta(t_1-t_2) - H_R(t_1,t_2) = \sum_{k=1}^{\infty} u_k(t_1)\,u_k^*(t_2) - \sum_{k=1}^{\infty} \lambda_k\,u_k(t_1)\,u_k^*(t_2)
$$
$$
= \sum_{k=1}^{\infty} (1-\lambda_k)\,u_k(t_1)\,u_k^*(t_2) \,, \tag{4.14}
$$

where we have used that the inverse Weyl transform of 1 is $\delta(t_1-t_2)$, that the eigensignals $u_k(t)$ of R are an orthonormal basis of $\mathcal{L}_2(\mathbb{R})$ and hence $\delta(t_1-t_2) = \sum_{k=1}^{\infty} u_k(t_1)\,u_k^*(t_2)$, and that $H_R(t_1,t_2) = \sum_{k=1}^{\infty} \lambda_k\,u_k(t_1)\,u_k^*(t_2)$ according to (4.3). We conclude from (4.14) that the eigensignals $u_k(t)$ of R are also the eigensignals of $\bar{R}$, and that the eigenvalues of $\bar{R}$ are given by

$$\bar{\lambda}_k = 1 - \lambda_k \,.$$

The eigenspace $\mathcal{U}_R$ is spanned by the first N_R eigensignals $u_k(t)$, which are exactly the eigensignals for which $\lambda_k > 1/2$,

$$\mathcal{U}_R = \mathrm{span}\{u_k(t)\}_{k=1}^{N_R} \,.$$

Since the entire set $\{u_k(t)\}_{k=1}^{\infty}$ is an orthonormal basis of $\mathcal{L}_2(\mathbb{R})$, the remaining eigensignals $u_k(t)$ span the orthogonal complement space $\overline{\mathcal{U}_R}$ of $\mathcal{U}_R$,

$$\overline{\mathcal{U}_R} = \mathrm{span}\{u_k(t)\}_{k=N_R+1}^{\infty} \,. \tag{4.15}$$

The eigenvalues associated to these remaining eigensignals satisfy $\lambda_k \le 1/2$ and, assuming that $1/2$ is not an eigenvalue, $\lambda_k < 1/2$. This implies $\bar{\lambda}_k = 1 - \lambda_k > 1/2$ for $k \ge N_R + 1$. Since the eigenvalues of $\bar{R}$ are $\bar{\lambda}_k = 1 - \lambda_k$, the eigenspace of $\bar{R}$ is spanned by all eigensignals for which $\bar{\lambda}_k > 1/2$, so that

$$\mathcal{U}_{\bar{R}} = \mathrm{span}\{u_k(t)\}_{k=N_R+1}^{\infty} \,. \tag{4.16}$$

Combining (4.15) and (4.16) gives $\mathcal{U}_{\bar{R}} = \overline{\mathcal{U}_R}$ as claimed. $\square$

In the theoretical case that $1/2$ happens to be an eigenvalue of R, it is still true that the eigenspaces of R and $\bar{R}$ are orthogonal, but the sum of these eigenspaces is no longer the entire space $\mathcal{L}_2(\mathbb{R})$. Instead, we have $\mathcal{U}_R + \mathcal{U}_{\bar{R}} = \mathcal{L}_2(\mathbb{R}) \setminus \mathcal{U}_\Delta$ where $\mathcal{U}_\Delta$ is the space spanned by all eigensignals whose corresponding eigenvalues are $1/2$.

We emphasize that the complement property of eigenspaces discussed above cannot be extended to a partition of the TF plane into more than two regions (see Section 5.3.1).

4.3 Subspace-Constrained Optimum Time-Frequency Synthesis

The synthesis method proposed in Section 4.1 and further discussed in Section 4.2 allows the construction of the optimum TF space (optimum in the sense of minimum TF localization error) for any given TF region R. A variation of this synthesis method, presented in this section, is motivated by the desire to enforce certain additional properties of the TF space. For example, we might desire the synthesized space to be strictly band-limited to a given frequency band. (Of course, such a band-limitation constraint will normally make sense only if the TF region R is not located entirely outside the given band.) This band-limitation property and some other important properties can be mathematically phrased as *subspace constraints*, i.e., the TF space $\mathcal{X}$ to be synthesized is constrained to be a subspace of a given linear signal space (the *constraint space*) $\mathcal{C}$. For example, this constraint space may be the space of all signals band-limited to a given frequency band.

Adopting the MLE criterion, the TF localization error $\epsilon(\mathcal{X}, R)$ has now to be minimized under the side constraint $\mathcal{X} \subseteq \mathcal{C}$. This leads to the following *subspace-constrained synthesis problem*:

$$\mathcal{X}^{(\mathcal{C})}_{R,\mathrm{MLE}} \;\triangleq\; \arg \min_{\mathcal{X} \subseteq \mathcal{C}} \; \epsilon(\mathcal{X}, R) \;=\; \arg \min_{\mathcal{X} \subseteq \mathcal{C}} \; \| I_R - W_{\mathcal{X}} \| \,.$$

Adopting the MRC criterion, the optimum MRC subspace of given dimension N (with $N \leq N_{\mathcal{C}}$) is defined as

$$\mathcal{X}^{(\mathcal{C})}_{R,\mathrm{MRC}}(N) \;\triangleq\; \arg \max_{\substack{N_{\mathcal{X}} = N \\ \mathcal{X} \subseteq \mathcal{C}}} \; \rho(\mathcal{X}, R) \;=\; \arg \max_{\substack{N_{\mathcal{X}} = N \\ \mathcal{X} \subseteq \mathcal{C}}} \; \langle W_{\mathcal{X}}, I_R \rangle \,.$$

From (3.20), it again follows that for prescribed (i.e., fixed) dimension of the space, $N_{\mathcal{X}} = N$, minimization of the TF localization error is equivalent to maximization of the regional TF concentration,

$$\mathcal{X}^{(\mathcal{C})}_{R,\mathrm{MLE}}(N) \;=\; \mathcal{X}^{(\mathcal{C})}_{R,\mathrm{MRC}}(N) \,.$$

Proceeding as in Section 2.7.2, we first introduce the *$\mathcal{C}$-restricted operator*

$$\mathbf{H}_{R,\mathcal{C}} \;\triangleq\; \mathbf{P}_{\mathcal{C}} \, \mathbf{H}_R \, \mathbf{P}_{\mathcal{C}} \,,$$

where $\mathbf{H}_R$ has been defined in (4.2) and $\mathbf{P}_{\mathcal{C}}$ is the orthogonal projection operator on the constraint space $\mathcal{C}$. The eigenvalues μ_k and eigenfunctions $v_k(t)$ of $\mathbf{H}_{R,\mathcal{C}}$ (again arranged in non-increasing order, $\mu_1 \geq \mu_2 \geq \ldots$) will be called the *$\mathcal{C}$-restricted eigenvalues* and *$\mathcal{C}$-restricted eigensignals*, respectively, of the TF region R. Furthermore, the *$\mathcal{C}$-restricted, N-dimensional eigenspace* $\mathcal{V}^{(N)}_R$ of R

(with $N \leq N_C$) is defined as the space spanned by the first N C-restricted eigensignals $v_k(t)$ of R, i.e., the N eigensignals with largest eigenvalues μ_k,

$$\mathcal{V}_R^{(N)} \triangleq \text{span}\{v_k(t)\}_{k=1}^N, \qquad N \leq N_C.$$

From Theorem 2.2 (see Section 2.7.2), it follows that the C-restricted eigenspaces provide the solution to the subspace-constrained synthesis problem:

Theorem 4.6 *The N-dimensional signal space minimizing the TF localization error $\epsilon(\mathcal{X}, R)$ and maximizing the regional TF concentration $\rho(\mathcal{X}, R)$ under the subspace constraint $\mathcal{X} \subseteq C$ is the C-restricted, N-dimensional eigenspace of R,*

$$\mathcal{X}_{R,\text{MLE}}^{(C)}(N) = \mathcal{X}_{R,\text{MRC}}^{(C)}(N) = \mathcal{V}_R^{(N)}, \qquad N \leq N_C.$$

The resulting (minimum) TF localization error is related to the eigenvalues λ_k of R and the C-restricted eigenvalues μ_k of R as

$$\epsilon^2(\mathcal{X}, R)\Big|_{\min,C,N} = \epsilon^2(\mathcal{X}, R)\Big|_{\mathcal{X}=\mathcal{V}_R^{(N)}} = \epsilon_0^2 + \sum_{k=1}^N (1-\mu_k)^2 + \sum_{k=N+1}^{N_C} \mu_k^2, \quad (4.17)$$

where

$$\epsilon_0^2 = A_R - A_{R,C} \geq 0 \quad \text{with} \quad A_R = \sum_{k=1}^{\infty} \lambda_k^2, \quad A_{R,C} \triangleq \sum_{k=1}^{N_C} \mu_k^2. \qquad (4.18)$$

The resulting (maximum) regional TF concentration is the arithmetic mean of the N first (largest) C-restricted eigenvalues of R,

$$\rho(\mathcal{X}, R)\Big|_{\max,C,N} = \rho(\mathcal{X}, R)\Big|_{\mathcal{X}=\mathcal{V}_R^{(N)}} = \frac{1}{N} \sum_{k=1}^N \mu_k. \qquad (4.19)$$

Proof. Since $N_{\mathcal{X}} = N$ is fixed, the subspace-constrained maximization of $\rho(\mathcal{X}, R)$ is equivalent to the maximization of the spaces' regional energy $E_{\mathcal{X}}^{(R)} = \langle W_{\mathcal{X}}, I_R \rangle$ subject to $\mathcal{X} \subseteq C$. Hence, Theorem 2.2 is directly applicable if we set $G(t, f) = I_R(t, f)$ and $\mathbf{H} = \mathbf{H}_R$. This shows that the N-dimensional space maximizing $E_{\mathcal{X}}^{(R)}$ is spanned by the N first (i.e., largest) eigenfunctions of $\mathbf{H}_{R,C}$. Thus, the optimum space is $\mathcal{V}_R^{(N)}$. It also follows from Theorem 2.2 that the regional energy achieved becomes $E_{\mathcal{X}}^{(R)} = \sum_{k=1}^N \mu_k$, which yields (4.19). Finally, (4.17) is derived from (4.19) as follows. Using (3.20), and subtracting and adding $A_{R,C} \triangleq \sum_{k=1}^{N_C} \mu_k^2$, we obtain

$$\epsilon^2(\mathcal{X}, R)\Big|_{\mathrm{min},\mathcal{C},N} = A_R - A_{R,\mathcal{C}} + A_{R,\mathcal{C}} + N - 2N\rho(\mathcal{X}, R)\Big|_{\mathrm{max},\mathcal{C},N}$$

$$= \epsilon_0^2 + A_{R,\mathcal{C}} + N - 2N\rho(\mathcal{X}, R)\Big|_{\mathrm{max},\mathcal{C},N}$$

where[1] $\epsilon_0^2 = A_R - A_{R,\mathcal{C}} = \sum_{k=1}^{\infty} \lambda_k^2 - \sum_{k=1}^{N_{\mathcal{C}}} \mu_k^2 \geq 0$ (cf. (4.9)). With (4.19), the second error component $A_{R,\mathcal{C}} + N - 2N\rho(\mathcal{X}, R)\big|_{\mathrm{max},\mathcal{C},N}$ is easily formulated as $\sum_{k=1}^{N}(1 - \mu_k)^2 + \sum_{k=N+1}^{N_{\mathcal{C}}} \mu_k^2.$ $\qquad\square$

We note that the error component ϵ_0 in (4.17), (4.18) is independent of N, and will be zero if $\mathcal{C} = \mathcal{L}_2(\mathbb{R})$ (i.e., in the case of no subspace constraint).

So far, we have prescribed the space dimension N. However, with (4.17) and the fact that ϵ_0 is independent of N, it is readily seen that the optimum value of N equals the number of all $\mathcal{C}$-restricted eigenvalues μ_k larger than $1/2$:

Corollary 4.2 *The signal space minimizing the TF localization error $\epsilon(\mathcal{X}, R)$ under the subspace constraint $\mathcal{X} \subseteq \mathcal{C}$ is the $\mathcal{C}$-restricted eigenspace of R with dimension $N_{R,\mathcal{C}}$,*

$$\mathcal{X}_{R,\mathrm{MLE}}^{(\mathcal{C})} = \mathcal{V}_R^{(N_{R,\mathcal{C}})},$$

where $N_{R,\mathcal{C}}$ denotes the number of $\mathcal{C}$-restricted eigenvalues μ_k larger than $1/2$. The residual (absolutely minimum) TF localization error is

$$\epsilon^2(\mathcal{X}, R)\Big|_{\mathrm{min},\mathcal{C}} = \epsilon^2(\mathcal{X}, R)\Big|_{\mathcal{X}=\mathcal{V}_R^{(N_{R,\mathcal{C}})}} = \epsilon_0^2 + \sum_{k=1}^{N_{R,\mathcal{C}}}(1 - \mu_k)^2 + \sum_{k=N_{R,\mathcal{C}}+1}^{N_{\mathcal{C}}} \mu_k^2.$$

In the case where the dimension of the constraint subspace $\mathcal{C}$ is small and an (arbitrary) orthonormal basis $\{c_j(t)\}_{j=1}^{N_{\mathcal{C}}}$ of $\mathcal{C}$ is available, the following method for calculating the $\mathcal{C}$-restricted eigenspace $\mathcal{V}_R^{(N)}$ is advantageous. This method is a straightforward application of Theorem 2.3.

1. Form the Hermitian $N_{\mathcal{C}} \times N_{\mathcal{C}}$ matrix $\mathbf{\Gamma}$ with elements

$$(\mathbf{\Gamma})_{ij} = \langle I_R, W_{c_i,c_j} \rangle = \langle \mathbf{H}_R c_j, c_i \rangle, \qquad i,j = 1,...,N_{\mathcal{C}}.$$

2. Calculate the eigenvalues μ_k and the normalized eigenvectors $\mathbf{v}_k$ of $\mathbf{\Gamma}$. (The eigenvalues μ_k equal the $\mathcal{C}$-restricted eigenvalues of R, and are assumed to be arranged in non-increasing order.)

---

[1] We note that $A_R = \sum_{k=1}^{\infty} \lambda_k^2$ and $A_{R,\mathcal{C}} = \sum_{k=1}^{N_{\mathcal{C}}} \mu_k^2$ equal the Hilbert-Schmidt norms [Naylor and Sell, 1982] of $\mathbf{H}_R$ and $\mathbf{H}_{R,\mathcal{C}}$, respectively. Since the Hilbert-Schmidt norm of $\mathbf{H}_{R,\mathcal{C}}$ can be shown to be not larger than that of $\mathbf{H}_R$, it follows that $A_R \geq A_{R,\mathcal{C}}$.

3. The $\mathcal{C}$-restricted eigensignals $v_k(t)$ spanning $\mathcal{V}_R^{(N)}$ are given as

$$v_k(t) \;=\; \sum_{j=1}^{N_C} \alpha_{kj}\, c_j(t)\,, \qquad k = 1, ..., N\,,$$

where the coefficient α_{kj} is the jth element of the kth eigenvector $\mathbf{v}_k$,

$$\alpha_{kj} \;=\; (\mathbf{v}_k)_j\,, \qquad k = 1, ..., N\,, \quad j = 1, ..., N_C\,.$$

4.4   Discrete-Time Synthesis Algorithms

The synthesis methods presented in Sections 4.1 and 4.3 must be reformulated in a discrete-time setting if they are to be implemented on a digital computer. The synthesis of a discrete-time signal space $\mathcal{X}$ is based on the discrete-time WD $W_{\mathcal{X}}^{(D)}(n, \theta)$ introduced in Section 2.6.2 (see (2.54), (2.55)).

4.4.1   The Discrete-Time Synthesis Problem

While the formulation of the discrete-time synthesis problem is initially analogous to that of the continuous-time synthesis problem, the discrete-time situation is somewhat more complicated due to the aliasing problem inherent in the discrete-time WD. In fact, we recall from Section 2.6.2 that $W_{\mathcal{X}}^{(D)}(n, \theta)$ will contain aliasing unless $\mathcal{X} \subseteq \mathcal{H}^{(\theta_0)}$. Here, $\mathcal{H}^{(\theta_0)}$ denotes the halfband subspace of all signals band-limited to a halfband defined by $|\theta - \theta_0| < 1/4$, where the halfband's center frequency θ_0 is arbitrary but fixed.

Let R be some "region" of the "(n, θ)-plane," i.e., some set of points (n, θ) with n integer and $|\theta - \theta_0| < 1/4$. Furthermore, let the indicator function $I_R(n, \theta)$ be 1 for $(n, \theta) \in R$ and 0 for $(n, \theta) \notin R$. In principle, the discrete-time synthesis problem is the calculation of the discrete-time signal space $\mathcal{X}_{R,\mathrm{MLE}}$ minimizing the TF localization error $\epsilon(\mathcal{X}, R)$ defined as

$$\epsilon^2(\mathcal{X}, R) \;=\; \left\| I_R - W_{\mathcal{X}}^{(D)} \right\|^2 \;=\; \sum_n \int_{\theta_0 - 1/4}^{\theta_0 + 1/4} \left[I_R(n, \theta) - W_{\mathcal{X}}^{(D)}(n, \theta) \right]^2 d\theta\,.$$

However, in order to avoid aliasing (which would cause the TF localization error $\epsilon(\mathcal{X}, R)$ to be meaningless), we include the *halfband constraint* $\mathcal{X} \subseteq \mathcal{H}^{(\theta_0)}$ in our synthesis problem, and define the optimum space to be

$$\mathcal{X}_{R,\mathrm{MLE}} \;\overset{\triangle}{=}\; \arg \min_{\mathcal{X} \subseteq \mathcal{H}^{(\theta_0)}} \epsilon(\mathcal{X}, R) \;=\; \arg \min_{\mathcal{X} \subseteq \mathcal{H}^{(\theta_0)}} \left\| I_R - W_{\mathcal{X}}^{(D)} \right\|\,. \tag{4.20}$$

This can be generalized by constraining the space $\mathcal{X}$ to be a subspace of some given discrete-time signal space $\mathcal{C} \subseteq \mathcal{H}^{(\theta_0)}$. This leads to the following *subspace-constrained synthesis problem*:

$$\mathcal{X}_{R,\text{MLE}}^{(\mathcal{C})} \triangleq \arg\min_{\mathcal{X} \subseteq \mathcal{C}} \epsilon(\mathcal{X}, R) = \arg\min_{\mathcal{X} \subseteq \mathcal{C}} \left\| I_R - W_{\mathcal{X}}^{(D)} \right\| \quad \text{where} \quad \mathcal{C} \subseteq \mathcal{H}^{(\theta_0)}. \tag{4.21}$$

Since $\mathcal{C} \subseteq \mathcal{H}^{(\theta_0)}$, aliasing is again avoided.

The discrete-time synthesis algorithms described below will make use of the fact that any halfband signal $x(n) \in \mathcal{H}^{(\theta_0)}$ is uniquely specified by, e.g., the even-indexed samples $x(2\nu)$ since the odd-indexed samples can be derived from the even-indexed samples by means of the interpolation

$$x(2\nu + 1) = 2 \sum_{\nu'} h\big(2(\nu - \nu') + 1\big)\, x(2\nu')\,. \tag{4.22}$$

Here,

$$h(n) = \frac{1}{2}\operatorname{sinc}\left(\pi \frac{n}{2}\right) e^{j2\pi\theta_0 n} \tag{4.23}$$

is the impulse response of an idealized halfband filter with center frequency θ_0.

For the sake of brevity, we shall present algorithms directly for the general subspace-constrained synthesis problem in (4.21), and consider the halfband-constrained synthesis problem in (4.20) as a special case. The algorithms summarized in the following two subsections generalize the discrete-time signal synthesis algorithms derived in [Hlawatsch and Krattenthaler, 1997] (see also [Hlawatsch and Kozek, 1994, Hlawatsch and Kozek, 1995]).

4.4.2   Basis Method

We first formulate a synthesis algorithm which assumes that an orthonormal basis $\{c_j(n)\}_{j=1}^{N_c}$ of $\mathcal{C}$ is available. This algorithm is particularly advantageous if the dimension of the constraint subspace $\mathcal{C}$ is small.

1. The indicator function $I_R(n, \theta)$ is projected onto the induced WD-domain subspace corresponding to the halfband subspace $\mathcal{H}^{(\theta_0)}$; this projection amounts to the convolution [Hlawatsch and Krattenthaler, 1997, Hlawatsch and Kozek, 1994, Hlawatsch and Kozek, 1995]

$$\tilde{I}_R(n, \theta) = \sum_{n'} W_h(n - n', \theta)\, I_R(n', \theta) \qquad \text{for} \quad |\theta - \theta_0| < 1/4 \tag{4.24}$$

where

$$W_h(n, \theta) = \big(1 - 4|\theta - \theta_0|\big) \operatorname{sinc}\big[\pi(1 - 4|\theta - \theta_0|)\, n\big] \qquad \text{for} \quad |\theta - \theta_0| < 1/4$$

is the discrete-time WD of the halfband-filter impulse response $h(n)$ [Claasen and Mecklenbräuker, 1980a]. This convolution is seen to be a lowpass filtering with respect to n, where the lowpass filter's bandwidth is $(1 - 4|\theta - \theta_0|)/2$ and thus depends on θ.

2. The projected indicator function $\tilde{I}_R(n, \theta)$ is transformed as

$$\tilde{H}_R(2\nu_1, 2\nu_2) = \int_{\theta_0 - 1/4}^{\theta_0 + 1/4} \tilde{I}_R(\nu_1 + \nu_2, \theta)\, e^{j4\pi(\nu_1 - \nu_2)\theta}\, d\theta \,. \tag{4.25}$$

3. The Hermitian $N_C \times N_C$ matrix $\mathbf{\Gamma}$ with elements

$$(\mathbf{\Gamma})_{ij} = 4 \sum_{\nu_1} \sum_{\nu_2} \tilde{H}_R(2\nu_1, 2\nu_2)\, c_i^*(2\nu_1)\, c_j(2\nu_2) \tag{4.26}$$

is calculated.

4. The eigenvalues μ_k and the normalized eigenvectors $\mathbf{v}_k$ of $\mathbf{\Gamma}$ are computed. (The eigenvalues μ_k are assumed to be arranged in non-increasing order.)

5. The optimum space is

$$\mathcal{X}_{R,\text{MLE}}^{(\mathcal{C})} = \text{span}\{v_k(n)\}_{k=1}^{N_{R,C}} \,;$$

it is spanned by the orthonormal basis functions

$$v_k(n) = \sum_{j=1}^{N_C} \alpha_{kj}\, c_j(n) \,, \qquad k = 1, ..., N \,,$$

where the coefficient α_{kj} is the jth element of the kth eigenvector $\mathbf{v}_k$,

$$\alpha_{kj} = (\mathbf{v}_k)_j \,, \qquad k = 1, ..., N \,, \quad j = 1, ..., N_C \,,$$

and the optimum space dimension $N_{R,C}$ is the number of eigenvalues μ_k larger than $1/2$.

The halfband-constrained synthesis problem (4.20) corresponds to the simple special case $\mathcal{C} = \mathcal{H}^{(\theta_0)}$. Here, the projection step (4.26) can be omitted. In fact, if the orthonormal basis spanning $\mathcal{C} = \mathcal{H}^{(\theta_0)}$ is chosen as $c_j(n) = \sqrt{2}\, h(n - 2j)$ (with $h(n)$ defined in (4.23)), then there is directly $(\mathbf{\Gamma})_{ij} = 2\, \tilde{H}_R(2i, 2j)$.

In practice, the WD is discretized also with respect to the frequency variable θ. Both the lowpass filtering (4.24) and the inverse Fourier transforms (4.25) can then be performed efficiently by means of FFT techniques. We note that a suboptimum, reduced-cost algorithm is obtained by omitting the initial projection step (4.24); the resulting space will then still be a subspace of the constraint space $\mathcal{C}$ but it will no longer minimize the synthesis error as postulated in (4.21). However, experiments have shown that the difference between this suboptimum space and the optimum solution is typically not dramatic.

4.4.3 Basis-Free Method

A second version of the discrete-time synthesis method does not require an orthonormal basis of the constraint subspace $\mathcal{C}$; instead, it assumes that $\mathcal{C}$ is characterized by its orthogonal projection operator. The first two steps of this "basis-free" method are identical with the first two steps of the basis method, but the remaining steps are different:

1. The orthogonal projection of the indicator function $I_R(n,\theta)$ onto the induced TF-domain halfband space is calculated according to (4.24).

2. The indicator function $\tilde{I}_R(n,\theta)$ is transformed as in (4.25).

3. The function $\tilde{H}_R(2\nu_1, 2\nu_2)$ is projected according to

$$\tilde{H}_{R,\mathcal{C}}(2\nu_1, 2\nu_2) \;=\; 4 \sum_{\nu_1'} \sum_{\nu_2'} P_\mathcal{C}(2\nu_1, 2\nu_1')\, P_\mathcal{C}^*(2\nu_2, 2\nu_2')\, \tilde{H}_R(2\nu_1', 2\nu_2') \;, \quad (4.27)$$

   where $P_\mathcal{C}(n_1, n_2)$ is the kernel of the orthogonal projection operator on $\mathcal{C}$.

4. The eigenvalues μ_k and eigenfunctions $w_k(\nu)$ of $\tilde{H}_{R,\mathcal{C}}(2\nu_1, 2\nu_2)$ are found by solving the eigenproblem

$$\sum_{\nu_2} \tilde{H}_{R,\mathcal{C}}(2\nu_1, 2\nu_2)\, w_k(\nu_2) \;=\; \mu_k\, w_k(\nu_1) \;.$$

   (The eigenvalues μ_k are assumed to be arranged in non-increasing order.)

5. The optimum space is

$$\mathcal{X}_{R,\mathrm{MLE}}^{(\mathcal{C})} \;=\; \mathrm{span}\{v_k(n)\}_{k=1}^{N_{R,\mathcal{C}}} \;;$$

   it is spanned by orthonormal basis functions $v_k(n)$ that are derived from the eigenfunctions $w_k(\nu)$ through an interpolation by a factor of 2 (see (4.22)),

$$v_k(2\nu) \;=\; w_k(\nu)\,, \qquad v_k(2\nu+1) \;=\; 2\sum_{\nu'} h\big(2(\nu-\nu')+1\big)\, w_k(2\nu')\,.$$

   The optimum space dimension $N_{R,\mathcal{C}}$ is the number of eigenvalues μ_k larger than $1/2$.

All these discrete-time operations can be formulated in a compact matrix-vector notation. As in the basis method, the first two steps can be implemented efficiently by means of FFT techniques. Again, a suboptimum, reduced-cost algorithm is obtained by omitting the initial projection step.

In the case of the halfband-constrained synthesis problem (4.20), i.e., $\mathcal{C} = \mathcal{H}^{(\theta_0)}$, the projection (4.27) can be omitted since $\tilde{H}_{R,\mathcal{C}}(2\nu_1, 2\nu_2) = \tilde{H}_R(2\nu_1, 2\nu_2)$.

4.5 Simulation Results

This section presents computer simulation results illustrating the performance of TF synthesis. The suboptimum, reduced-cost, discrete-time synthesis algorithms (without the initial halfband projection step) were used in all examples.

Fig. 4.1 shows the synthesis of a space localized in a C-shaped TF region. The region and its eigenvalues are depicted in Fig. 4.1(a) and (b), respectively. Since 20 eigenvalues are above $1/2$, the space's dimension is 20. The WD of the synthesized space and the WD of one of the space's basis signals are shown in Fig. 4.1(c) and (d), respectively. Whereas the WDs of the basis signals contain large inner interference (cross) terms [Hlawatsch and Flandrin, 1997], such interference terms are quite small in the WD of the overall space; thus, the space is *simple* (cf. Section 3.1.2). A similar effect was observed in Fig. 2.3.

Fig. 4.2 shows an interesting threshold effect of TF synthesis. The indicator function $I_R(n, \theta)$ was replaced by a more general "model" function $M(n, \theta)$; the optimum space is here defined as $\mathcal{X}_{\text{opt}} = \arg\min_{\mathcal{X} \subseteq \mathcal{C}} \left\| M - W_{\mathcal{X}}^{(D)} \right\|$. As shown in Fig. 4.2(a), $M(n, \theta)$ assumes the heights 0, 0.3 and 0.7 in disjoint subregions. A similar height distribution is observed in the eigenvalues shown in Fig. 4.2(b). The eigenfunctions corresponding to the 8 eigenvalues that are larger than $1/2$ are TF located inside the subregion corresponding to model height 0.7, as is evident from Fig. 4.2(c). In general, the synthesized space will be located in those TF regions where the model function $M(n, \theta)$ is larger than $1/2$; all other model components will be disregarded by TF synthesis.

Fig. 4.3 gives an example of subspace-constrained TF synthesis. The TF region is a slanted strip region. The constraint subspace is the space of all signals band-limited to a band with bandwidth $1/4$, i.e., one half the WD's frequency period $1/2$. From Fig. 4.3(c), we verify that the space is properly band-limited (apart from small errors caused by the finite time support of all signals calculated). The dimension of the optimum space is $N_{R,\mathcal{C}} = 9$.

4.6 Hermite Spaces and Prolate Spheroidal Spaces Reconsidered

According to Sections 3.3.2 and 3.4.2, the *Hermite spaces* and the *prolate spheroidal spaces* have optimum TF concentration. We shall now show that these spaces are also optimum from the viewpoint of TF synthesis, i.e., they minimize the TF localization error with respect to suitably defined TF regions.

4.6.1 Hermite Spaces

The next theorem, shown in Appendix 4.A, states that the Hermite spaces $\mathcal{H}_N^{(T)}$ are the optimum TF spaces (eigenspaces) of an *elliptical* TF region.

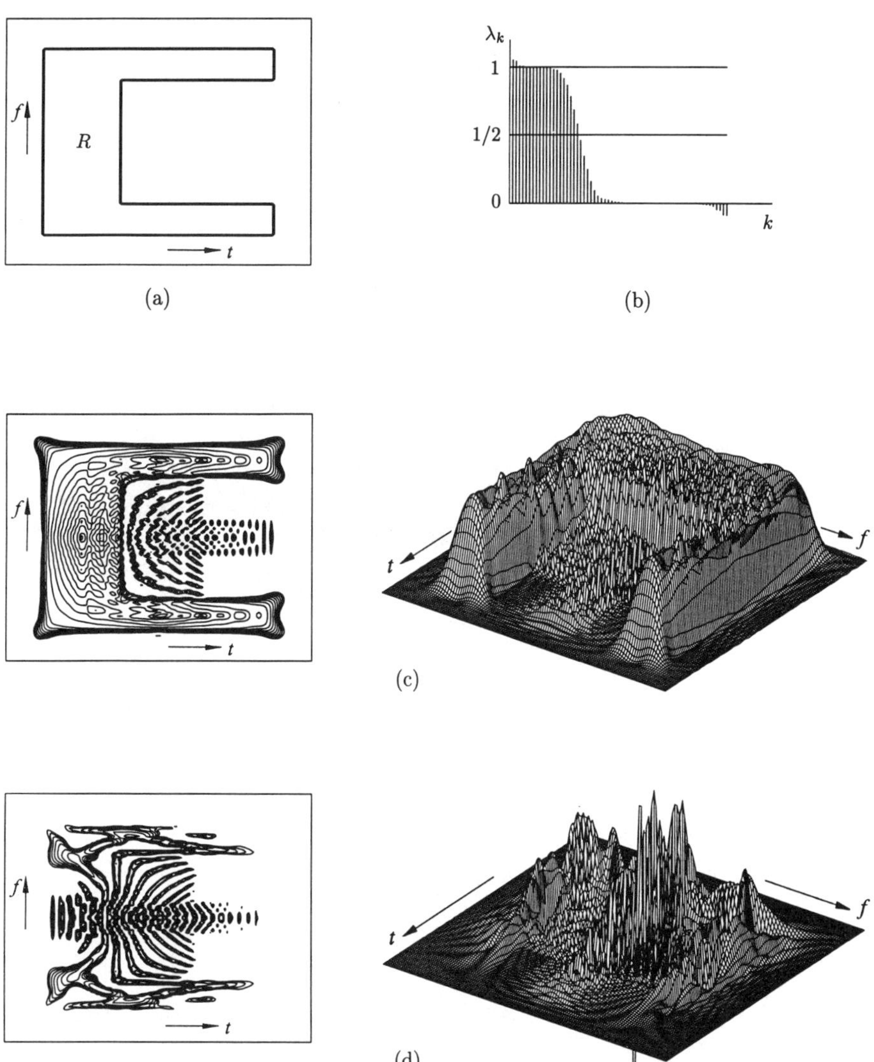

Figure 4.1. TF synthesis of a space localized in a C-shaped TF region: (a) TF region, (b) eigenvalues, (c) WD of synthesized space, (d) WD of a basis signal.

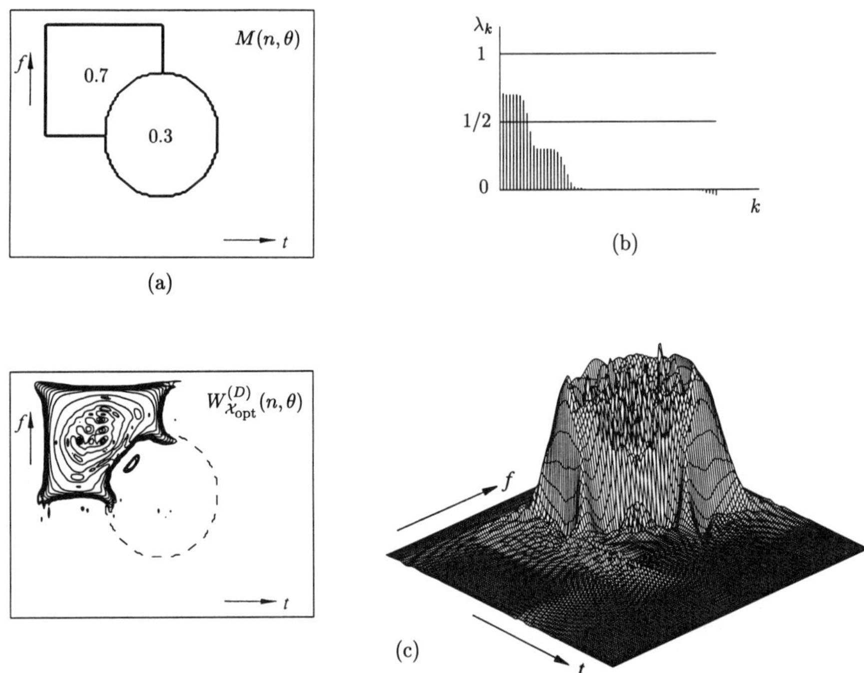

Figure 4.2. Threshold effect of TF synthesis: (a) Model function, (b) eigenvalues, (c) WD of synthesized space.

Theorem 4.7 *Let the TF region R be the ellipse defined by $(t/T)^2 + (Tf)^2 \leq K^2$. Then, the eigenspace of R (i.e., the space minimizing the TF localization error $\epsilon(\mathcal{X}, R)$) is the Hermite space of dimension N_R,*

$$\mathcal{X}_{R,\mathrm{MLE}} = \mathcal{U}_R = \mathcal{H}_{N_R}^{(T)},$$

where the optimum dimension N_R (the number of eigenvalues of R that are $> 1/2$) depends only on K, i.e., N_R is independent of T.

Equivalently, N_R depends only on the *area* $A_R = \pi K^2$ of the elliptical TF region R. This dependence was analyzed numerically by performing discrete-time synthesis from various elliptical TF regions with integer-valued areas between 1 and 50, and determining the number of eigenvalues $> 1/2$. We obtained

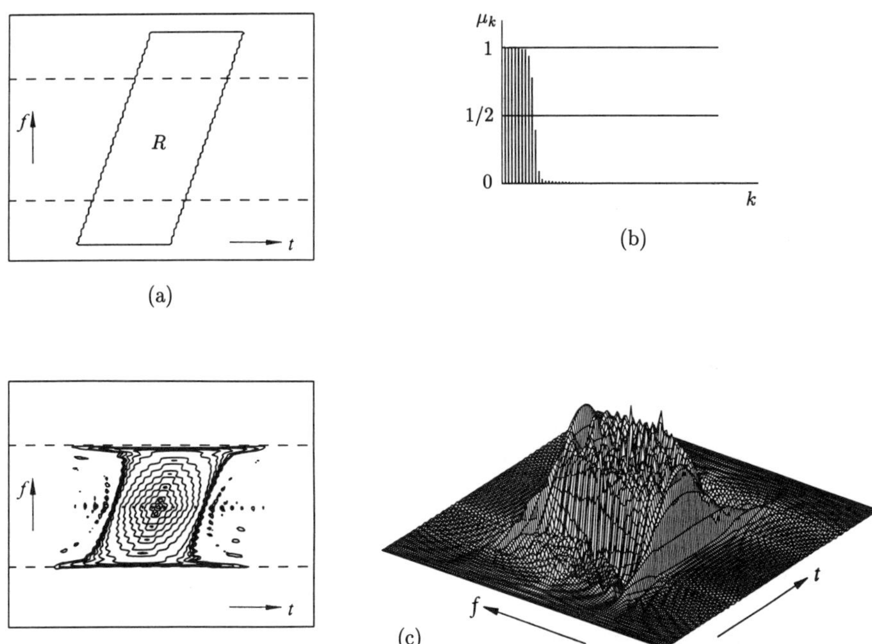

Figure 4.3. Subspace-constrained (band-limited) TF synthesis of a space localized in a slanted strip region: (a) TF region and constraint band (shown by broken lines), (b) subspace-restricted eigenvalues (only positive values shown), (c) WD of synthesized space.

$N_R = A_R$ for $A_R = 1, 2, ..., 7$ but $N_R = A_R + 1$ for $A_R = 8, 9, ..., 50$. This deviation between N_R and A_R is consistent with the asymptotic result in Eq. (2.83) of [Duijvelaar, 1984]. In fact, from this asymptotic result it can be concluded that, for elliptical TF regions with area A_R, the optimum space dimension is $N_R = A_R + K(A_R)$ where $K(A_R)$ is on the order of $A_R^{1/3}$ [Janssen, 1997a]. Note, however, that the *relative* deviation from A_R, i.e., $(N_R - A_R)/A_R$, tends to zero for $A_R \to \infty$. We also note that an expression for the eigenvalues of an elliptical region is given in Eq. (4.A.4) in Appendix 4.A.

The eigenspaces of more general elliptical regions can be constructed by combining Theorems 4.7 and 4.4. Indeed, any general elliptical TF region $\tilde{R}$ can be obtained from a special elliptical region R: $(t/T)^2 + (Tf)^2 \leq K^2$ by means of an area-preserving coordinate transform $(t, f) \to (\alpha t + \beta f - \tau, \gamma t + \delta f - \nu)$. With Theorem 4.4, the eigenspaces of $\tilde{R}$ are then obtained as $\mathcal{U}_{\tilde{R}}^{(N)} = \mathbf{H}\, \mathcal{H}_N^{(T)}$, where $\mathbf{H}$ is the unitary signal transformation corresponding to the above TF coordinate transform in the sense of Theorem 4.4.

4.6.2 Prolate Spheroidal Spaces

Similar to the Hermite spaces, the prolate spheroidal spaces $\mathcal{P}_N^{(T,F)}$ can be viewed as optimum TF spaces, i.e., as spaces minimizing the TF localization error with respect to a suitably defined TF region. However, in this case we have to include a band-limitation side constraint. The next theorem is proved in Appendix 4.B.

Theorem 4.8 *Let the TF region R be the infinite strip $[-T/2, T/2] \times (-\infty, \infty)$ comprising all TF points (t, f) with $t \in [-T/2, T/2]$. Then, the $\mathcal{F}[-F/2, F/2]$-restricted eigenspace of R (i.e., the space minimizing the TF localization error $\epsilon(\mathcal{X}, R)$ under the band-limitation constraint $\mathcal{X} \subseteq \mathcal{F}[-F/2, F/2]$) is the prolate spheroidal space of dimension $N_{R,\mathcal{F}}$,*

$$\mathcal{X}_{R,\mathrm{MLE}}^{(\mathcal{F}[-F/2, F/2])} = \mathcal{P}_{N_{R,\mathcal{F}}}^{(T,F)},$$

where the optimum dimension $N_{R,\mathcal{F}}$ (the number of $\mathcal{F}[-F/2, F/2]$-restricted eigenvalues of R that are $> 1/2$) depends only on the product TF.

4.7 Extension to Other Space Representations

Thus far, the TF synthesis of signal spaces has been based on the optimum approximation of the indicator function of a TF region R by the WD of a linear signal space. This basic approach can be extended by replacing the WD of a space with some other quadratic space representation (cf. Section 2.2). Our synthesis methods can be reformulated for any other quadratic space representation provided that this representation is *unitary*, i.e., the corresponding quadratic signal representation satisfies Moyal's formula [Hlawatsch, 1992a, Hlawatsch and Krattenthaler, 1992]. An example is the family of *generalized WDs* [Janssen, 1982, Hlawatsch, 1992a, Kozek, 1992a], although there is little motivation for replacing the WD by some other member of this family [Janssen, 1982, Kozek and Hlawatsch, 1992, Hlawatsch and Urbanke, 1994, Hlawatsch and Flandrin, 1997]. Of more practical interest are the *ambiguity function* discussed in Chapter 7 or some unitary time-frequency or time-scale representation such as the *Bertrand P_0-distribution*, the *Altes-Marinovich Q-distribution*, or the *power-WD* [Rioul and Flandrin, 1992, Bertrand and Bertrand, 1992a, Bertrand and Bertrand, 1992b, Flandrin and Gonçalvès, 1994, Flandrin and Gonçalvès, 1996, Hlawatsch et al., 1993b, Altes, 1990, Marinovich, 1986, Papandreou et al., 1993, Hlawatsch et al., 1997, Hlawatsch et al., 1993a, Papandreou et al., 1995].

Appendix 4.A: Proof of Theorem 4.7

The indicator function of the elliptical TF region R: $\left(\frac{t}{T}\right)^2 + (Tf)^2 \leq K^2$ can be written as

$$I_R(t,f) = \tilde{w}_K\left(\left(\frac{t}{T}\right)^2 + (Tf)^2\right) \qquad \text{with} \quad \tilde{w}_K(\beta) = \begin{cases} 1, \ 0 \leq \beta \leq K^2 \\ 0, \ \beta > K^2 \end{cases}$$

$$(4.A.1)$$

(note that $\tilde{w}_K(\beta)$ is defined for $\beta \geq 0$ only). Now, any square-integrable function defined for $\beta \geq 0$ can be expanded into the orthonormal basis [Abramowitz and Stegun, 1965, Wilcox, 1991] $l_n(\beta) \triangleq \sqrt{4\pi}\, L_n(4\pi\beta)\, e^{-2\pi\beta}$ ($\beta \geq 0$, $n = 0, 1, ...$), where $L_n(\beta) = \frac{1}{n!} e^\beta \frac{d^n}{d\beta^n}\left(\beta^n e^{-\beta}\right)$ ($n = 0, 1, ...$) is the Laguerre polynomial of order n [Abramowitz and Stegun, 1965]. Hence, we can write $\tilde{w}_K(\beta) = \sum_{n=0}^\infty r_n\, l_n(\beta)$ where

$$r_n = \langle \tilde{w}_K, l_n \rangle = \sqrt{4\pi} \int_0^{K^2} L_n(4\pi\beta)\, e^{-2\pi\beta}\, d\beta. \qquad (4.A.2)$$

With (4.A.1) and using (2.44), (2.45), we obtain

$$\begin{aligned}
I_R(t,f) &= \sum_{n=0}^\infty r_n\, l_n\left(\left(\frac{t}{T}\right)^2 + (Tf)^2\right) \\
&= \sum_{n=0}^\infty r_n\, \sqrt{4\pi}\, L_n\left(4\pi\left[\left(\frac{t}{T}\right)^2 + (Tf)^2\right]\right) e^{-2\pi\left[\left(\frac{t}{T}\right)^2 + (Tf)^2\right]} \\
&= \sum_{k=1}^\infty \lambda_k\, W_{h_k^{(T)}}(t,f)
\end{aligned}$$

$$(4.A.3)$$

where

$$\lambda_k = \sqrt{\pi}\, (-1)^{k-1}\, r_{k-1} = \frac{1}{2}(-1)^{k-1} \int_0^{4\pi K^2} L_{k-1}(\beta)\, e^{-\beta/2}\, d\beta \qquad (4.A.4)$$

for $k = 1, 2, ...$ (we note that $\lambda_k \leq 1 - e^{-2\pi K^2}$, see Section 4.6 in [Janssen, 1997b]). Since the λ_k are real-valued and the Hermite functions $h_k^{(T)}(t)$ are orthonormal, (4.A.3) is recognized to be the eigenexpansion (4.4) of $I_R(t,f)$. Hence, the eigensignals of $I_R(t,f)$ are the Hermite functions $h_k^{(T)}(t)$, and consequently the eigenspaces of $I_R(t,f)$ are the Hermite spaces $\mathcal{H}_N^{(T)}$. Finally, since the number N_R of eigenvalues $\lambda_k = \sqrt{\pi}\, (-1)^{k-1}\, r_{k-1}$ larger than $1/2$ depends only on the expansion coefficients r_n of $\tilde{w}_K(\beta)$, which according to (4.A.2) depend only on K, it is clear that N_R depends only on K.

Appendix 4.B: Proof of Theorem 4.8

In what follows, let $\mathcal{T} \triangleq \mathcal{T}[-T/2, T/2]$ and $\mathcal{F} \triangleq \mathcal{F}[-F/2, F/2]$ for brevity. The indicator function of the TF region $R = [-T/2, T/2] \times (-\infty, \infty)$ is

$$I_R(t, f) = \text{rect}_T(t) = \begin{cases} 1, & |t| \leq T/2 \\ 0, & |t| > T/2 \,, \end{cases}$$

and it is easily shown that the corresponding operator $\mathbf{H}_R$ has the kernel

$$H_R(t_1, t_2) = \text{rect}_T\left(\frac{t_1 + t_2}{2}\right) \delta(t_1 - t_2) \,. \tag{4.B.1}$$

For prescribed dimension $N_{\mathcal{X}} = N$, the synthesis problem considered can be phrased as the maximization of the regional TF concentration

$$\rho(\mathcal{X}, R) = \frac{1}{N_{\mathcal{X}}} \langle W_{\mathcal{X}}, I_R \rangle = \frac{\int_{-T/2}^{T/2} \int_f W_{\mathcal{X}}(t, f) \, dt \, df}{\int_t \int_f W_{\mathcal{X}}(t, f) \, dt \, df}$$

subject to the subspace constraint $\mathcal{X} \subseteq \mathcal{F}$. Comparing with (3.14), we see that the regional TF concentration $\rho(\mathcal{X}, R)$ here reduces to the temporal concentration $\alpha_{\mathcal{X}}^{(T)}$ defined in Section 3.4.1,

$$\rho(\mathcal{X}, R) = \alpha_{\mathcal{X}}^{(T)} \,.$$

According to Theorem 3.2 (see Section 3.4.2), $\alpha_{\mathcal{X}}^{(T)}$ is maximized (subject to $\mathcal{X} \subseteq \mathcal{F}$) by the prolate spheroidal spaces $\mathcal{P}_N^{(T, F)}$.

It is instructive to work out the $\mathcal{F}$-restricted operator $\mathbf{H}_{R, \mathcal{F}}$. With (4.B.1) and the kernel of $\mathbf{P}_{\mathcal{F}}$ being $P_{\mathcal{F}}(t_1, t_2) = F \, \text{sinc}[\pi F(t_1 - t_2)]$, we can show

$$\mathbf{H}_{R, \mathcal{F}} = \mathbf{P}_{\mathcal{F}} \mathbf{P}_{\mathcal{T}} \mathbf{P}_{\mathcal{F}} = [\mathbf{P}_{\mathcal{T}}]_{\mathcal{F}} \,.$$

The eigenfunctions of $[\mathbf{P}_{\mathcal{T}}]_{\mathcal{F}}$ have been shown in Appendix 3.B to equal the prolate spheroidal wave functions.

We now consider the optimum dimension. According to Corollary 4.2 (see Section 4.3), the optimum dimension $N_{R, \mathcal{F}}$ (optimum in the sense of minimum localization error) is the number of $\mathcal{F}$-restricted eigenvalues of R that are larger than $1/2$, and thus $N_{R, \mathcal{F}}$ depends only on these eigenvalues. The eigenvalues of $\mathbf{H}_{R, \mathcal{F}} = [\mathbf{P}_{\mathcal{T}}]_{\mathcal{F}}$ equal the eigenvalues $\lambda_k^{(T, F)}$ of the operator $\mathbf{K}^{(T, F)}$ (see Appendix 3.B). It is well known [Papoulis, 1984b] that the eigenvalues $\lambda_k^{(T, F)}$ depend only on the product TF. Hence, also $N_{R, \mathcal{F}}$ depends only on TF.

5   TIME-FREQUENCY FILTERS AND TIME-FREQUENCY EXPANSIONS

The separation of signal components occupying effectively disjoint regions of the TF plane is a fundamental problem of TF signal processing that is encountered in many applications. A related problem is the parsimonious representation (expansion) of signals located in a given TF region. In this chapter, we propose solutions to both problems. These solutions are based on the optimum space synthesis (eigenspaces) described in the previous chapter. A "TF projection filter" for signal separation is obtained as the orthogonal projection operator on the eigenspace of the given TF pass region, and a "TF expansion" is based on any orthonormal basis of the eigenspace of the given TF support region.

The basic TF projection filter can be extended to *TF filter banks* satisfying the perfect reconstruction property. These TF filter banks correspond to a partition of the TF plane, with each subregion of the TF partition being the TF pass region of one of the filters.

This chapter is organized as follows. Section 5.1 gives some motivation and background for the TF filter and TF expansion problems. Section 5.2 discusses "TF projections," i.e., orthogonal projections onto eigenspaces, which provide the solution to both the TF filter and TF expansion problems. Furthermore, simulation results demonstrating the good performance of TF projection filters

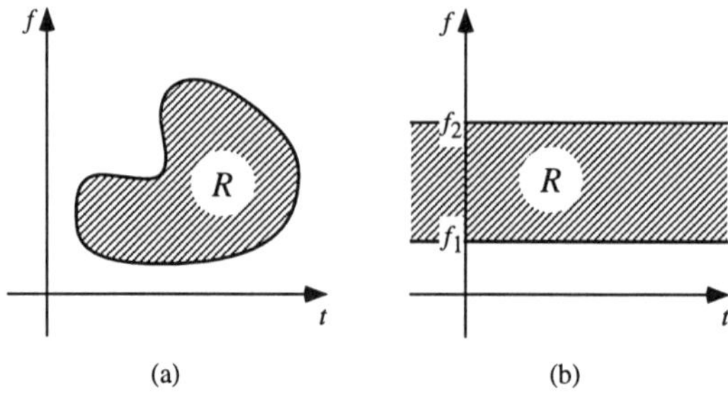

Figure 5.1.   Time-frequency pass region: (a) general, (b) corresponding to frequency band.

are presented. Section 5.3 proposes and studies perfect-reconstruction TF filter banks based on subspace-constrained TF synthesis. Finally, possible extensions and variations of TF projection filters are briefly pointed out in Section 5.4.

5.1 Motivation and Background

Often, it is desirable to filter a signal contaminated by some interfering signal (e.g. noise), or to expand a signal into an orthonormal basis such that a minimum number of expansion coefficients are required. Appropriate solutions to these two problems depend on the signal model and the prior knowledge available. We shall here assume that prior knowledge about the signal's *TF support*, i.e., the effective support of the signal's WD, is available[1]. If the signal is a (generally nonstationary) random process, then its TF support is defined as the effective support of its Wigner-Ville spectrum (expected WD) [Martin and Flandrin, 1985, Flandrin, 1989, Flandrin and Martin, 1997].

The two problems considered in this chapter are stated as follows:

- *TF filtering* — the construction of a filter with given "TF pass region" R (see part (a) of **Fig. 5.1**), i.e., a filter that passes all signals located inside[2] R but suppresses all signals located outside R.

- *TF expansion* — the construction of an orthonormal basis for the parsimonious expansion of all signals located in a given TF region R. The expansion coefficients represent the signal and can be used for further signal processing.

---

[1]This requires careful distinction between WD "signal terms" and WD "interference terms" [Hlawatsch and Flandrin, 1997]. By definition, the interference terms are not part of the TF support as they do not contain signal energy.

[2]A signal will be considered to be inside a TF region R if its *effective* TF support (defined above) is contained in R. We emphasize that no signal can be exactly contained in a bounded TF region.

5.1.1 A Classical Example and Its Generalization

As a motivation for our approach to solve the TF filtering and TF expansion problems, we first assume that the joint TF localization is replaced by a pure frequency localization, i.e., the TF pass or support region R is formally replaced by a frequency interval (band) $[f_1, f_2]$. This is a special case of the situation considered previously: the TF region R now is an infinite strip running parallel to the time axis, $R = (-\infty, \infty) \times [f_1, f_2]$ (see Fig. 5.1(b)). Here, the theoretically appropriate solution to the filter problem is simply an idealized bandpass filter, i.e., a linear, time-invariant filter with frequency response

$$H(f) = \begin{cases} 1, & f \in [f_1, f_2] \\ 0, & f \notin [f_1, f_2]. \end{cases}$$

Furthermore, an appropriate solution to the expansion problem is any orthonormal basis spanning the linear space $\mathcal{F}[f_1, f_2]$ of all signals band-limited to $[f_1, f_2]$—for example, the basis of appropriately scaled and shifted sinc functions, or the basis of appropriately frequency-shifted prolate spheroidal wave functions [Slepian and Pollak, 1961, Landau and Pollak, 1961, Papoulis, 1984b]. The two solutions are closely related, since the idealized bandpass filter with passband $[f_1, f_2]$ is the *orthogonal projection operator* on the linear space $\mathcal{F}[f_1, f_2]$ of all signals band-limited to $[f_1, f_2]$ [Naylor and Sell, 1982]. *Thus, the signal space $\mathcal{F}[f_1, f_2]$ provides the solution to both the TF filtering and TF expansion problems.* Note that, according to Section 2.4, $\mathcal{F}[f_1, f_2]$ is ideally localized in the TF region R since its WD is 1 inside R and 0 outside R.

Let us now return to a general TF region R as shown in Fig. 5.1(a). With the conceptual background developed above, we are able to formulate a unified approach to the solution of the TF filtering and TF expansion problems:

- We construct the linear signal space with optimum TF localization in the TF region R, i.e., the eigenspace $\mathcal{U}_R$ of R (see Section 4.1).

- The *TF filter* is the orthogonal projection operator on the eigenspace $\mathcal{U}_R$. This "TF projection filter" [Hlawatsch and Kozek, 1994] is a linear, generally time-varying system.

- The basis used for the *TF expansion* is any orthonormal basis spanning the eigenspace $\mathcal{U}_R$.

Thus, the problems of TF filtering and TF expansion have been reduced to the optimum synthesis of TF spaces, a problem whose solution has been discussed in detail in Chapter 4.

5.1.2   Time-Frequency Filters

The TF projection filter proposed above is a linear, generally time-varying system. Linearity is obviously desirable, and the time-varying nature is dictated by the general shape[3] of the TF pass region R. The orthogonal projection structure implies that the filter is a self-adjoint, idempotent linear operator. This is a natural property since the filtering task is to *pass* signals in some TF region and *reject* signals in the rest of the TF plane. The orthogonal projection structure here follows from two requirements:

- The filter's output signal (if nonzero at all) is always located inside the TF pass region R. Thus, if it is passed once again through the filter, it should not be changed any more. This requires that the linear operator be *idempotent*, i.e., a projection [Naylor and Sell, 1982].

- The part of the input signal $x(t)$ that is passed by the filter $\mathbf{H}$, i.e., the output signal $(\mathbf{H}x)(t)$, and the part that is rejected, $x(t) - (\mathbf{H}x)(t)$, are approximately TF disjoint. It follows from Moyal's formula (1.10) that two strictly TF disjoint signals are orthogonal. Orthogonality of $(\mathbf{H}x)(t)$ and $x(t) - (\mathbf{H}x)(t)$ requires that the projection operator be *orthogonal* [Naylor and Sell, 1982].

5.1.3   Related Work

Various schemes for TF filtering (also called "TF localization" in the mathematical literature) have been proposed previously (see [Kozek and Hlawatsch, 1992] for a comparative discussion). A conceptually simple method consists of a masking of the signal's WD followed by signal synthesis [Boudreaux-Bartels and Parks, 1986, Boudreaux-Bartels, 1997, Hlawatsch and Krattenthaler, 1997] and results in a highly nonlinear overall filter whose performance has been shown to be potentially poor [Krattenthaler and Hlawatsch, 1993, Kozek and Hlawatsch, 1992, Hlawatsch et al., 1994]. Replacing the WD by a linear TF representation (such as the short-time Fourier transform [Portnoff, 1980, Daubechies, 1988, Bourdier et al., 1988], the Gabor expansion [Farkash and Raz, 1994], or the wavelet transform [Daubechies and Paul, 1988]) results in a linear filter whose performance is often satisfactory but is influenced by the window or wavelet used, and is also restricted by a TF resolution tradeoff [Kozek and Hlawatsch, 1992]. Other linear TF filter designs are based on the *Weyl symbol* [Kozek, 1992b, Kozek and Hlawatsch, 1991a, Kozek and Hlawatsch, 1992] or the *WD of a linear system* [Hlawatsch, 1992b]. Although a quan-

---

[3]A linear time-*in*variant filter would only allow TF pass regions consisting of strips as in Fig. 5.1(b).

titative assessment of performance is difficult due to the uncertainty principle (specifically, due to the fact that no signal may be exactly contained in a
bounded TF region), the performance of the TF projection filter proposed here
was generally observed in simulation studies to be as good as, or better than,
the performance of other TF filtering methods [Kozek and Hlawatsch, 1992].

A classical method for the construction of TF subspaces and TF expansions
is based on the prolate spheroidal wave functions [Slepian and Pollak, 1961,
Landau and Pollak, 1961, Papoulis, 1984b]. The underlying TF regions are
here restricted to rectangular shapes. A mathematical operator framework of
TF concentrated basis systems has been introduced in [Parks and Shenoy, 1990],
however without an explicit method for constructing the relevant operator for
a given TF region. Finally, it is clear that any set of TF concentrated functions
which are "sufficiently dense" in a given TF region can be used as a (generally
nonorthogonal) basis of a TF subspace. This includes sets of Gabor logons
[Parks and Shenoy, 1990, Umesh and Tufts, 1992] or wavelet functions. This
approach, however, has certain drawbacks due to the limited concentration of
the functions used and the necessity of orthogonalizing the set of functions.

5.2   Time-Frequency Projections

The TF projection filters and TF signal expansions proposed here are both
based on the orthogonal projection onto the TF region's eigenspace $\mathcal{U}_R$. We
shall now investigate some aspects of this "TF projection," and we shall present
simulation results illustrating its performance.

5.2.1   Time-Frequency Projection Filters

With (1.1) and (1.2), the orthogonal projection of a signal $x(t)$ onto $\mathcal{U}_R$ is

$$(\mathbf{P}_{\mathcal{U}_R} x)(t) \;=\; \int_{t'} P_{\mathcal{U}_R}(t, t')\, x(t')\, dt' \tag{5.1}$$

$$=\; \sum_{k=1}^{N_R} \langle x, u_k \rangle\, u_k(t) \,, \tag{5.2}$$

where $P_{\mathcal{U}_R}(t, t')$ denotes the kernel of the orthogonal projection operator $\mathbf{P}_{\mathcal{U}_R}$
on $\mathcal{U}_R$, and the $u_k(t)$ are the eigensignals of R (which form an orthonormal basis
of $\mathcal{U}_R$). Expression (5.1) is the input-output relation of a *linear, time-varying
system (filter)* with impulse response $P_{\mathcal{U}_R}(t, t')$. We call this filter the *TF projection filter with pass region R*. Expression (5.2), which is particularly suited
to parallel processing, involves N_R inner products $\langle x, u_k \rangle = \int_t x(t)\, u_k^*(t)\, dt$ of
the input signal $x(t)$ with the first N_R eigensignals $u_k(t)$ of R. The calculation
of the eigensignals $u_k(t)$ has been discussed in Sections 4.1.2 and 4.4. The

computational expense associated with (5.2) is proportional to the subspace dimension N_R, and also increases (via the inner products $\langle x, u_k \rangle$) with the effective durations of the eigensignals $u_k(t)$. With $N_R \approx A_R$, it follows that *the computational expense of a TF projection filter increases with the area of the TF pass region.* The effective durations of the eigensignals depend primarily on the length of the TF region R in the time direction.

The TF projection filter $\mathbf{P}_{\mathcal{U}_R}$, which was defined above as the orthogonal projection operator on the space with optimum TF localization in the pass region R, can be derived using two alternative optimality criteria. First, $\mathbf{P}_{\mathcal{U}_R}$ is the linear, time-varying system $\mathbf{H}$ whose Weyl symbol $L_{\mathbf{H}}(t, f) = \int_\tau H\left(t + \frac{\tau}{2}, t - \frac{\tau}{2}\right) e^{-j2\pi f\tau} \, d\tau$ (cf. (2.58)) is closest to the TF indicator function $I_R(t, f)$, subject to the side constraint that the system $\mathbf{H}$ be an orthogonal projection operator [Kozek and Hlawatsch, 1992, Kozek, 1992b]. This formulation is possible since the WD of a signal space equals the Weyl symbol of the space's orthogonal projection operator (see (2.5)). Second, since the Weyl symbol of an orthogonal projection system also equals the system's *Wigner distribution* defined in [Hlawatsch, 1992b], this optimization can alternatively be formulated using the WD of a system instead of the system's Weyl symbol. Thus, $\mathbf{P}_{\mathcal{U}_R}$ is the linear, time-varying system $\mathbf{H}$ whose WD (as defined in [Hlawatsch, 1992b]) is closest to the TF indicator function $I_R(t, f)$, again under the side constraint that the system $\mathbf{H}$ be an orthogonal projection operator. Omitting the projection side constraint yields alternative filter designs (see Section 5.4).

5.2.2   Time-Frequency Signal Expansions

The *TF signal expansion* for given TF support region R amounts to forming the N_R expansion coefficients $\alpha_k = \langle x, u_k \rangle$ $(k = 1, ..., N_R)$. If the signal $x(t)$ is well concentrated in R (i.e., the regional TF concentration $\rho(x, R)$ defined in Section 3.5.3 is close to 1), then it will be represented with high accuracy by the coefficients α_k. These coefficients can then be used for further signal processing. We recall from Section 3.5.2 that the number of expansion coefficients, N_R, roughly equals the area A_R of the TF region R.

From the TF expansion coefficients $\alpha_k = \langle x, u_k \rangle$, the signal can be reconstructed according to

$$\hat{x}(t) = \sum_{k=1}^{N_R} \langle x, u_k \rangle u_k(t) = \left(\mathbf{P}_{\mathcal{U}_R} x\right)(t),$$

which yields the signal's orthogonal projection onto $\mathcal{U}_R$ and is thus equal to the output (5.1), (5.2) of the TF projection filter. The expansion/reconstruction error incurred, $\|x - \hat{x}\| = \|x - \mathbf{P}_{\mathcal{U}_R} x\|$, will be small for signals well concentrated in R. Often, we are also interested in rejecting any signal that is outside

R (e.g., to suppress noise or parasitic signal components, see below). The computational expense of a TF expansion is again determined by the dimension N_R (corresponding to the area A_R) and the effective eigensignal durations.

5.2.3 Pass/Reject Analysis

The "output" of both the TF projection filter and the TF signal expansion is the signal's orthogonal projection onto the eigenspace $\mathcal{U}_R$. We desire that this orthogonal projection passes signals located inside the TF pass region R but rejects signals located outside R. An obvious measure of how well an input signal $x(t)$ is passed by the orthogonal projection is the ratio[4] of the output energy $\|\mathbf{P}_{\mathcal{U}_R}x\|^2$ and the input energy $\|x\|^2$. According to Section 3.2.1, this ratio is just the affiliation of the input signal $x(t)$ to the eigenspace $\mathcal{U}_R$,

$$a(x|\mathcal{U}_R) \;=\; \frac{\|\mathbf{P}_{\mathcal{U}_R}x\|^2}{\|x\|^2}\;.$$

The affiliation $a(x|\mathcal{U}_R)$ should be nearly one for an input signal well inside the pass region R (i.e., when $\rho(x,R) \approx 1$) and nearly zero for an input signal well outside the pass region R (i.e., when $\rho(x,R) \approx 0$). Hence, the desired filter/expansion performance can be summarized as

$$a(x|\mathcal{U}_R) \;\approx\; \begin{cases} 1, & \text{for } x(t) \text{ such that } \rho(x,R) \approx 1, \\ 0, & \text{for } x(t) \text{ such that } \rho(x,R) \approx 0. \end{cases}$$

We recall from Section 3.5.3 that the affiliation inequality yields the approximate bounds

$$1-\epsilon_R \;\leq\; a(x|\mathcal{U}_R) \;\leq\; 1 \qquad \text{for} \quad \rho(x,R) \approx 1,$$
$$0 \;\leq\; a(x|\mathcal{U}_R) \;\leq\; \epsilon_R \qquad \text{for} \quad \rho(x,R) \approx 0,$$

which unfortunately are rather loose since the TF localization error ϵ_R is typically on the order of 1.

5.2.4 Noise Analysis

Often, the TF projection filter/TF expansion will either be used for suppressing additive noise, or noise suppression is at least a desired side effect. If $w(t)$ is zero-mean, stationary white noise with power spectral density $S_w(f) = \eta$, then the mean energy of the orthogonal projection $(\mathbf{P}_{\mathcal{U}_R}w)(t)$ is easily shown to be

---

[4]A related measure is the normalized deviation between the input signal and its orthogonal projection onto $\mathcal{U}_R$, $\bar{a}(x|\mathcal{U}_R) \overset{\triangle}{=} \|x - \mathbf{P}_{\mathcal{U}_R}x\|^2/\|x\|^2$. Using the idempotency of $\mathbf{P}_{\mathcal{U}_R}$, it is easily shown that $\bar{a}(x|\mathcal{U}_R) = 1 - a(x|\mathcal{U}_R)$.

$$E\{\|\mathbf{P}_{\mathcal{U}_R}w\|^2\} = \eta N_R \,, \tag{5.3}$$

which is proportional to the dimension N_R. Since $N_R \approx A_R$, we obtain the rule of thumb that *the mean noise energy passed is roughly proportional to the area of the TF pass region.*

If the input signal $x(t) = s(t) + w(t)$ consists of a deterministic signal component $s(t)$ with energy E_s and white noise $w(t)$ with power spectral density $S_w(f) = \eta$, the orthogonal projection of $x(t)$ onto $\mathcal{U}_R$ becomes $(\mathbf{P}_{\mathcal{U}_R}x)(t) = (\mathbf{P}_{\mathcal{U}_R}s)(t) + (\mathbf{P}_{\mathcal{U}_R}w)(t)$. We then define the *output signal-to-noise ratio* as the ratio of the energy $\|\mathbf{P}_{\mathcal{U}_R}s\|^2$ of the projected signal component and the mean energy $E\{\|\mathbf{P}_{\mathcal{U}_R}w\|^2\}$ of the projected noise. With $\|\mathbf{P}_{\mathcal{U}_R}s\|^2 = E_s\,a(s|\mathcal{U}_R)$ and $E\{\|\mathbf{P}_{\mathcal{U}_R}w\|^2\} = \eta N_R$, this output SNR is

$$\mathrm{SNR} \triangleq \frac{\|\mathbf{P}_{\mathcal{U}_R}s\|^2}{E\{\|\mathbf{P}_{\mathcal{U}_R}w\|^2\}} = \frac{E_s\,a(s|\mathcal{U}_R)}{\eta N_R} \,.$$

If $s(t)$ is well concentrated in R, there will be $(\mathbf{P}_{\mathcal{U}_R}s)(t) \approx s(t)$ and thus $a(s|\mathcal{U}_R) \approx 1$, so that

$$\mathrm{SNR} \approx \frac{E_s}{\eta N_R} \qquad \text{for } \rho(s,R) \approx 1 \,.$$

5.2.5   Simulation Results

We now present simulation results demonstrating the performance of TF projection filters. A TF filtering experiment is shown in **Fig. 5.2**. The three-component input signal consists of a Gaussian signal and two windowed quadratic FM[5] signals. The time duration of the overall signal is 128 samples. Since the signal components overlap with respect to both time and frequency, they cannot be separated using a simple time gating or a time-invariant filter.

In order to isolate the middle signal component, a TF pass region R was chosen as shown in Fig. 5.2(b), and the TF projection filter or, equivalently, the region's eigenspace was derived as detailed in Sections 4.1.2 and 4.4. According to Fig. 5.2(c), 6 eigenvalues are larger than $1/2$. Hence, the optimum space dimension is $N_R = 6$, which is close to the TF region's area $A_R = 5.3$. Fig. 5.2(d) shows the WD of the eigenspace $\mathcal{U}_R$. The residual TF localization error is $\epsilon_R = 0.87$. Fig. 5.2(e) depicts the output signal of the TF projection filter (i.e., the orthogonal projection of the three-component input signal onto the eigenspace $\mathcal{U}_R$). A comparison with the true signal component shown in Fig. 5.2(f) demonstrates that the filtering was successful in the sense that the

---

[5]The term "quadratic FM" refers to the signal's instantaneous frequency which is a quadratic function of time.

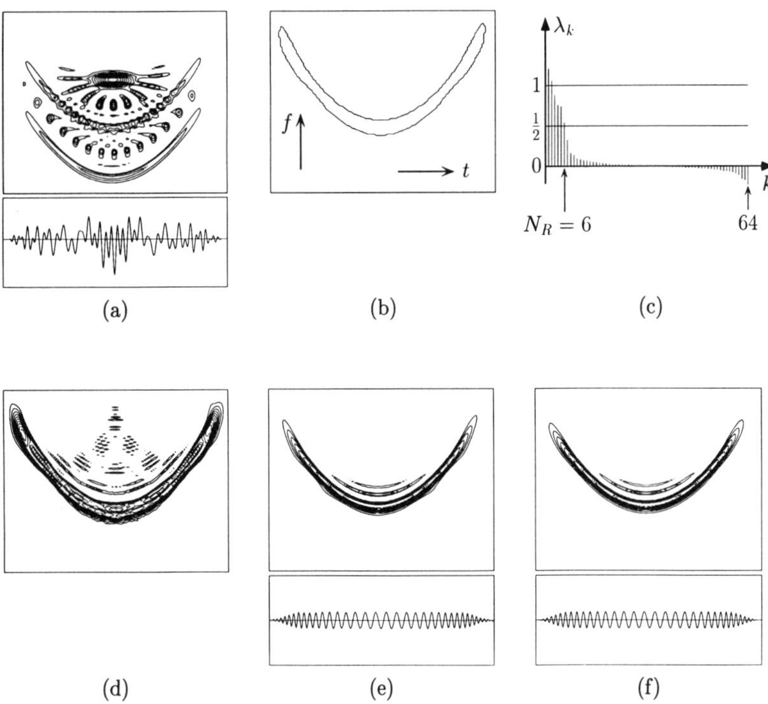

Figure 5.2. Application of TF projection filter to signal separation: (a) Real part and (slightly smoothed) WD of three-component input signal, (b) TF pass region R, (c) eigenvalues of TF region R, (d) WD of eigenspace $\mathcal{U}_R$, (e) output of TF projection filter (projection of input signal onto $\mathcal{U}_R$), (f) desired (true) signal component.

desired signal component is obtained with very little distortion while the other two components are well rejected. A quantitative performance characterization is given by the normalized output energies or affiliations of the individual signal components (see Section 5.2.3), which are $-0.02\,\text{dB}$ for the desired signal component and $-40.7\,\text{dB}$ and $-66.3\,\text{dB}$ for the two other signal components.

Fig. 5.3 illustrates the application of TF projection filters to noise suppression. The filter's input signal is the FM signal component in Fig. 5.2(f) contaminated by a realization of halfband-filtered[6] white noise with a signal-

---

[6]In order to avoid aliasing effects in the discrete-time WD [Claasen and Mecklenbräuker, 1980a], all signals used are restricted to one half of the total spectral period of discrete-time signals (cf. Sections 2.6.2 and 4.4).

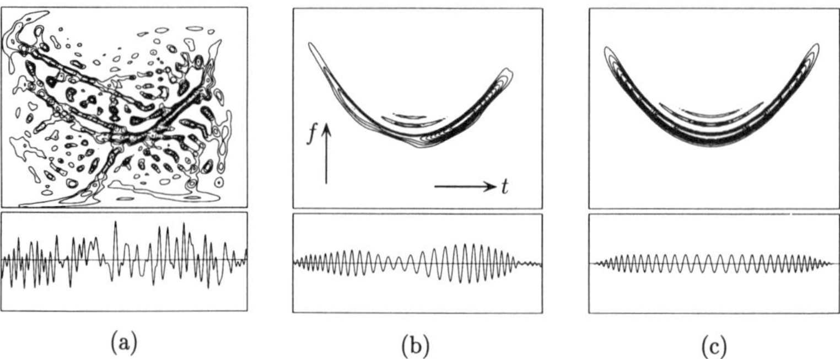

Figure 5.3. Application of TF projection filter to noise suppression: (a) Real part and (slightly smoothed) WD of noisy input signal, (b) output of TF projection filter, (c) desired (true) signal.

to-noise ratio (SNR) of $-3\,$dB. (The SNR is defined as the ratio of the energies of the FM signal and the noise signal.) The TF projection filter is the same as in the previous experiment. The output signal, shown in Fig. 5.3(b), is seen to be a reasonable estimate of the FM signal, apart from a parasitic amplitude modulation which is due to noise components located inside the pass region. The SNR of the output signal[7] is $6.9\,$dB. Hence, the overall SNR improvement achieved by the filter is $9.9\,$dB. Of course, this SNR improvement depends on the specific noise realization, which is random. In particular, the SNR improvement must be comparatively poor if a large part of the noise energy happens to fall inside the filter's pass region. Therefore, a more meaningful performance measure is the *mean* noise attenuation achieved by the filter (averaged over the entire noise ensemble), which is given by the ratio of the mean noise energies at the output and the input of the filter. It follows from (5.3) that the mean noise energy is proportional to the space dimension; hence, the noise attenuation factor is $6/64$ since the eigenspace dimension is $N_R = 6$ and the dimension of the "total signal space" (the space of all discrete-time halfband signals of length 128) is 64. This corresponds to a mean noise attenuation of $10.28\,$dB.

Our last experiment, depicted in **Fig. 5.4**, considers a noise suppression problem involving a natural signal, namely, two pitch periods of a voiced speech sound. Again, the signal length is 128 samples and the SNR of the input signal is $-3\,$dB. The TF pass region, shown in Fig. 5.4(b), was derived by thresholding

---

[7]This output SNR is based on the output signal's deviation from the true FM signal; it is different from the SNR considered in Section 5.2.4 in that (i) it is based on a single noise realization and not on the entire noise ensemble, and (ii) it reflects both the residual noise passed by the filter and the deterministic distortion of the FM signal caused by the filter.

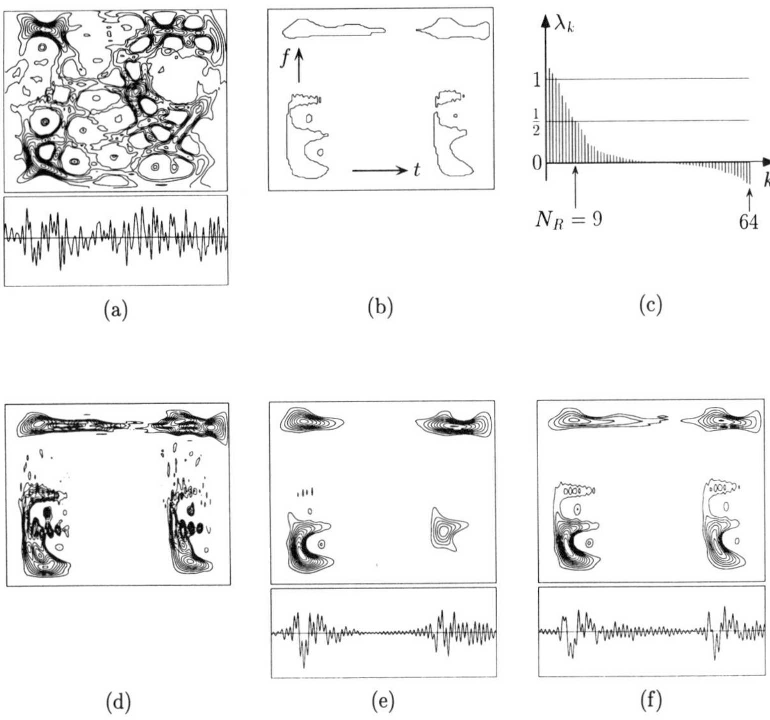

Figure 5.4. Application of TF projection filter to noise suppression involving a speech signal: (a) Real part and (slightly smoothed) WD of noisy speech signal, (b) TF pass region R, (c) eigenvalues of TF region R, (d) WD of eigenspace $\mathcal{U}_R$, (e) output of TF projection filter (projection of noisy speech signal onto $\mathcal{U}_R$), (f) desired (true) speech signal.

a smoothed WD of the noise-free speech signal. Note that the pass region is a multiple region. The area, dimension, and localization error are $A_R = 8.1$, $N_R = 9$, and $\epsilon_R = 1.4$, respectively. The SNR improvement achieved (defined as described above) is 7.4 dB. The mean noise attenuation is 9/64 or 8.52 dB.

The experiments presented are relevant not only to the performance of TF projection filters but also to that of TF signal expansions, since a TF signal expansion is based on the orthogonal projection onto the TF region's eigenspace. Specifically, the number of expansion coefficients required equals the dimension N_R, and the total mean noise energy contained in the expansion coefficients equals the mean noise energy passed by the orthogonal projection and is thus proportional to N_R. It follows from $N_R \approx A_R$ that, for a parsimonious signal

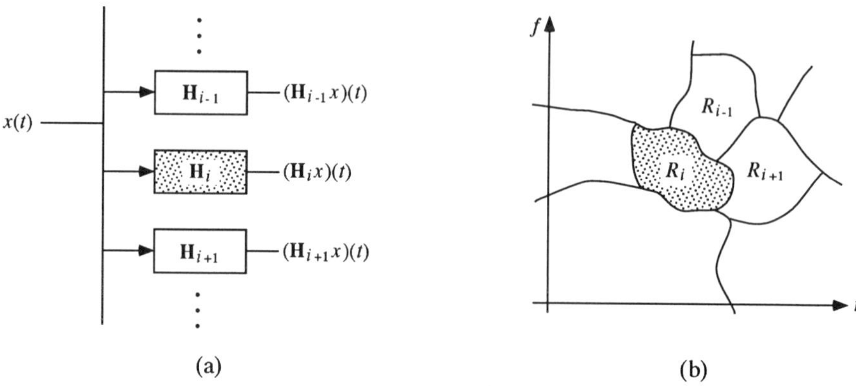

(a)   (b)

Figure 5.5.   (a) Filter bank, (b) corresponding partition of the TF plane.

expansion and good noise suppression, A_R should be as small as possible. This implies a rather precise prior knowledge of the signal's TF support.

The application of TF projection filters to the estimation and detection of noise-contaminated random signals will be considered in Chapter 6.

5.3   Time-Frequency Filter Banks

The concept of TF projection filters can be extended to the design of *TF filter banks* satisfying the property of perfect reconstruction [Kozek and Hlawatsch, 1991a].

5.3.1   Orthogonal Space Partitions

The problem considered in this section is the design of a filter bank corresponding to a partition of the TF plane (see **Fig. 5.5**). A partition of the TF plane is a set of M TF regions R_i which are disjoint (non-overlapping) and "complete" in the sense that the union of all regions R_i yields the entire TF plane,

$$R_i \cap R_j = \emptyset \quad \text{for } i \neq j \qquad \text{and} \qquad \bigcup_{i=1}^{M} R_i = \mathbb{R}^2 \, .$$

The TF filter bank is now defined conceptually as a set of M linear, time-varying filters $\mathbf{H}_i$, where the ith filter is supposed to pass all signals contained in the ith TF region R_i and to suppress all signals located outside R_i. Since the action of each $\mathbf{H}_i$ is to pass certain signals and suppress certain other signals, it is natural to construct $\mathbf{H}_i$ as an orthogonal projection operator on a suitably defined subspace $\mathcal{X}_i$, i.e., $\mathbf{H}_i = \mathbf{P}_{\mathcal{X}_i}$. We also require the overall filter bank to possess the *perfect reconstruction property* which postulates that the sum of all filter outputs $(\mathbf{P}_{\mathcal{X}_i} x)(t)$ equals the input signal $x(t)$ or, equivalently, that the sum of all $\mathbf{P}_{\mathcal{X}_i}$ equals the identity operator $\mathbf{I}$ on $\mathcal{L}_2(\mathbb{R})$,

$$\sum_{i=1}^{M} (\mathbf{P}_{\mathcal{X}_i} x)(t) = x(t) , \qquad \sum_{i=1}^{M} \mathbf{P}_{\mathcal{X}_i} = \mathbf{I} .$$

For the signal spaces $\mathcal{X}_i$, this implies

$$\mathcal{X}_i \perp \mathcal{X}_j \quad \text{for } i \neq j \qquad \text{and} \qquad \sum_{i=1}^{M} \mathcal{X}_i = \mathcal{L}_2(\mathbb{R}) . \tag{5.4}$$

i.e., the spaces $\mathcal{X}_i$ are an "orthogonal partition" of the total signal space $\mathcal{L}_2(\mathbb{R})$. Thus, by construction, the partition $\{R_i\}_{i=1}^{M}$ of the TF plane corresponds to a partition $\{\mathcal{X}_i\}_{i=1}^{M}$ of $\mathcal{L}_2(\mathbb{R})$.

We now have to find an optimum way of constructing a partition of $\mathcal{L}_2(\mathbb{R})$ into M orthogonal subspaces $\mathcal{X}_i$. A naive but conceptually simple approach to this problem is to choose the ith subspace $\mathcal{X}_i$ as the eigenspace of the ith TF region R_i, or equivalently, to choose the ith filter as the TF projection filter associated to R_i as discussed in Sections 5.1 and 5.2. However, we know from Section 4.2.4 that for $M > 2$ such an independent design of each filter will not result in an orthogonal partition of $\mathcal{L}_2(\mathbb{R})$, i.e., (5.4) and hence the perfect-reconstruction property will not be satisfied. It follows that the individual filters cannot be designed independently of each other if we want the perfect-reconstruction property of the overall filter bank as well as an orthogonal projector structure of each filter.[8]

From a theoretical viewpoint, it would be most satisfactory to formulate the design of all projection filters $\mathbf{P}_{\mathcal{X}_i}$ (or, equivalently, of all subspaces $\mathcal{X}_i$) as a joint optimization problem. A natural approach is to minimize the sum of the squared TF localization errors of all regions R_i,

$$\epsilon^2(\mathcal{X}_1, \mathcal{X}_2, ..., \mathcal{X}_M; R_1, R_2, ..., R_M) \stackrel{\triangle}{=} \sum_{i=1}^{M} \epsilon^2(\mathcal{X}_i, R_i) ,$$

where $\epsilon(\mathcal{X}_i, R_i) = \|I_{R_i} - W_{\mathcal{X}_i}\|$, under the side constraint (5.4). However, this minimization problem appears to be difficult to solve; furthermore, it must be expected that the method resulting from such an approach is computationally very expensive. We therefore propose two *recursive* design strategies which are suboptimum in the sense that they will not solve the above joint minimization

---

[8]An alternative design philosophy in which the individual filters are not constrained to be orthogonal projection operators is given by the *Weyl filter banks* introduced in [Kozek and Hlawatsch, 1991a].

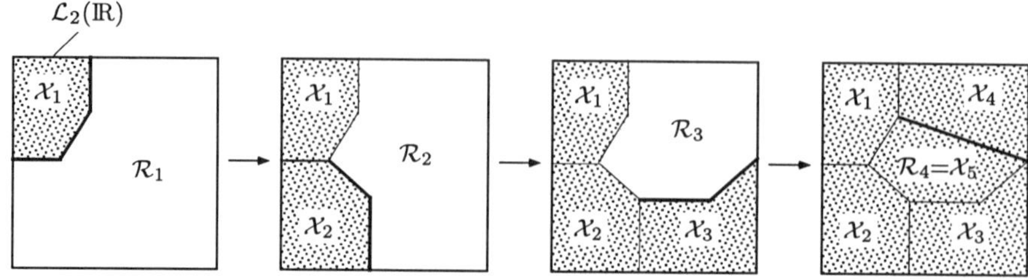

Figure 5.6.   Sequential space partitioning: example for $M = 5$.

problem, even though each step of the recursive design method is still based on an optimum procedure. These recursive methods make use of the subspace-constrained synthesis of spaces described in Sections 4.3 and 4.4. They both follow the idea of successively partitioning a space into two subspaces.

5.3.2 Sequential Partitioning

We first develop a *sequential* design strategy where the spaces $\mathcal{X}_i$ are synthesized one after the other. The synthesis of each space $\mathcal{X}_i$ is influenced by the corresponding TF region R_i and all spaces previously synthesized. This method assumes a specific ordering $R_1, R_2, ..., R_M$ of the TF regions; this ordering is of course arbitrary, but it will influence the resulting spaces $\mathcal{X}_i$ to a certain extent. The sequential method is summarized below; it is illustrated in **Fig. 5.6** for a specific example with $M = 5$ TF regions R_i.

1. The first space $\mathcal{X}_1$ is chosen as the space with optimum TF localization in the first TF region R_1, i.e., as the eigenspace $\mathcal{U}_{R_1}$ of R_1 (see Sections 4.1.2 and 4.4), except for the following potential difference: whereas the eigenspace $\mathcal{U}_{R_1}$ is spanned by all eigensignals $u_k(t)$ whose eigenvalues are larger than $1/2$, we here take the first (i.e., largest) $N_1 = \text{round}\{A_{R_1}\}$ eigenvalues (i.e., N_1 is the integer number that is closest to the area A_{R_1} of R_1).

    Then, the "residual space" $\mathcal{R}_1 = \overline{\mathcal{X}_1}$ is formed as the orthogonal complement of $\mathcal{X}_1$. Thus, at this point the total signal space $\mathcal{L}_2(\mathbb{R})$ has been partitioned into the space $\mathcal{X}_1$ corresponding to our first TF region R_1 and the residual space $\mathcal{R}_1$ orthogonal to $\mathcal{X}_1$, i.e., $\mathcal{R}_1 \perp \mathcal{X}_1$ and $\mathcal{X}_1 + \mathcal{R}_1 = \mathcal{L}_2(\mathbb{R})$.

2. The second space $\mathcal{X}_2$ is taken as the space with optimum TF localization in the second TF region R_2 under the subspace constraint $\mathcal{X}_2 \subseteq \mathcal{R}_1$, so that the space $\mathcal{X}_2$ is guaranteed to lie in the residual space $\mathcal{R}_1$. According to Section 4.3, $\mathcal{X}_2$ is the $\mathcal{R}_1$-restricted eigenspace of R_2. Again, we deviate from this solution in one respect: whereas the optimum space is spanned by the $\mathcal{R}_1$-

restricted eigensignals $v_k(t)$ corresponding to all $\mathcal{R}_1$-restricted eigenvalues μ_k that are larger than $1/2$, we here take the $\mathcal{R}_1$-restricted eigensignals corresponding to the $N_2 = \text{round}\,\{A_{R_2}\}$ largest μ_k. Note that, by construction, $\mathcal{X}_2$ is orthogonal to $\mathcal{X}_1$.

We now form the new residual space $\mathcal{R}_2 = \overline{\mathcal{X}_2} \cap \mathcal{R}_1$ as the orthogonal complement space of $\mathcal{X}_2$ in $\mathcal{R}_1$, i.e., as the space satisfying $\mathcal{R}_2 \perp \mathcal{X}_2$ and $\mathcal{X}_2 + \mathcal{R}_2 = \mathcal{R}_1$. Thus, the residual space $\mathcal{R}_1$ has been partitioned into the space $\mathcal{X}_2$ corresponding to R_2 and a new residual space $\mathcal{R}_2$ (usually smaller than $\mathcal{R}_1$) that is orthogonal to $\mathcal{X}_2$. Furthermore, at this point, the total signal space $\mathcal{L}_2(\mathbb{R})$ has been partitioned into the space $\mathcal{X}_1$ corresponding to R_1, the space $\mathcal{X}_2$ corresponding to R_2, and the new residual space $\mathcal{R}_2$ orthogonal to both $\mathcal{X}_1$ and $\mathcal{X}_2$.

3. The sequential synthesis method proceeds by repeating the last step. At the ith step, $\mathcal{X}_i$ is the space with optimum TF localization in R_i under the subspace constraint $\mathcal{X}_i \subseteq \mathcal{R}_{i-1}$, where $\mathcal{R}_{i-1}$ is the residual space obtained at the previous step. (Again, the synthesis of $\mathcal{X}_i$ is based on the $N_i = \text{round}\,\{A_{R_i}\}$ largest eigenvalues.) Then, the new residual space $\mathcal{R}_i = \overline{\mathcal{X}_i} \cap \mathcal{R}_{i-1}$ is constructed as the orthogonal complement of $\mathcal{X}_i$ in $\mathcal{R}_{i-1}$, i.e., $\mathcal{R}_i \perp \mathcal{X}_i$ and $\mathcal{X}_i + \mathcal{R}_i = \mathcal{R}_{i-1}$. Thus, after the ith step, $\mathcal{L}_2(\mathbb{R})$ has been partitioned into the *orthogonal* spaces $\mathcal{X}_1, \mathcal{X}_2, ..., \mathcal{X}_i$ corresponding to $R_1, R_2, ..., R_i$, respectively, plus a residual space $\mathcal{R}_i$ orthogonal to all spaces $\mathcal{X}_1$ through $\mathcal{X}_i$.

The TF projection filter bank obtained with this sequential partitioning method satisfies the perfect-reconstruction property since by construction the partition of the entire TF plane corresponds to a partition of $\mathcal{L}_2(\mathbb{R})$. Note that, in practice, the total signal space will not be $\mathcal{L}_2(\mathbb{R})$ but a space of discrete-time signals with finite length (i.e., a finite-dimensional space); similarly, the "entire TF plane" will have finite area. This makes the method practically realizable. The calculation of an orthonormal basis of the orthogonal complement spaces (residual spaces) $\mathcal{R}_i$ can be done by means of the Gram-Schmidt orthogonalization method [Naylor and Sell, 1982, Luenberger, 1969].

5.3.3 Binary Partitioning

An alternative recursive design method, which will be called *binary partitioning*, is also based on the idea of successively partitioning a space into two subspaces. However, in contrast to the "sequential" partitioning strategy discussed previously, these subspaces are initially not the looked-for spaces $\mathcal{X}_i$ but larger spaces. The binary partitioning strategy is summarized in the following; an example with $M = 5$ is shown in **Fig. 5.7**.

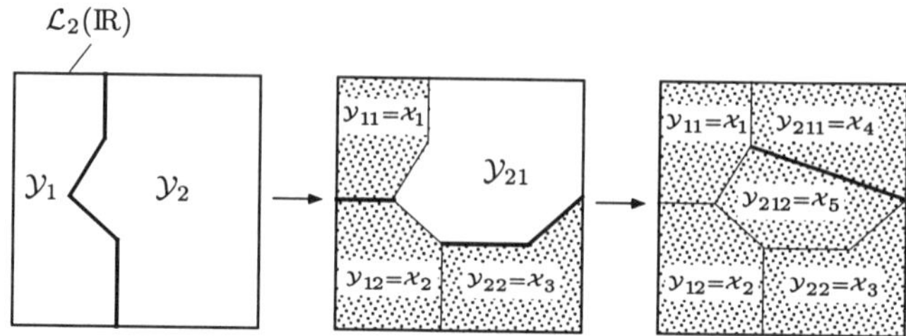

Figure 5.7.   Binary space partitioning: example for $M = 5$.

1. The TF regions R_i are grouped into two "super-regions" S_1 and S_2. Each super-region is the union of certain adjacent R_i; furthermore the super-regions are disjoint and their union is the entire TF plane.

   Next, the eigenspace of the first super-region S_1 is determined as discussed in Sections 4.1.2 and 4.4. This space, denoted as $\mathcal{Y}_1$, is hence $\mathcal{Y}_1 = \mathcal{U}_{S_1}$. The space $\mathcal{Y}_2$ (corresponding to the second super-region S_2) is taken to be the orthogonal complement space of $\mathcal{Y}_1$, i.e., $\mathcal{Y}_2 = \overline{\mathcal{Y}_1}$. Thus, the total space $\mathcal{L}_2(\mathrm{IR})$ has been partitioned into two orthogonal spaces $\mathcal{Y}_1$ and $\mathcal{Y}_2$, each of which corresponds to the union of certain adjacent TF regions R_i.

2. The TF regions contained in S_1 are grouped into two new super-regions S_{11} and S_{12}. These super-regions are disjoint and their union is S_1.

   Next, the space $\mathcal{Y}_{11}$ is defined as the $\mathcal{Y}_1$-restricted eigenspace of the first super-region S_{11} (see Sections 4.3 and 4.4). The space $\mathcal{Y}_{12}$ (corresponding to the second super-region S_{12}) is taken to be the orthogonal complement space of $\mathcal{Y}_{11}$ in $\mathcal{Y}_1$, i.e., $\mathcal{Y}_{12} = \overline{\mathcal{Y}_{11}} \cap \mathcal{Y}_1$ so that $\mathcal{Y}_{12} \perp \mathcal{Y}_{11}$ and $\mathcal{Y}_{11} + \mathcal{Y}_{12} = \mathcal{Y}_1$. Thus, the space $\mathcal{Y}_1$ has been partitioned into the orthogonal spaces $\mathcal{Y}_{11}$ and $\mathcal{Y}_{12}$, just as the super-region S_1 has been partitioned into S_{11} and S_{12}.

   In an analogous manner, the second super-region S_2 is partitioned into the regions S_{21} and S_{22}, and the "super-space" $\mathcal{Y}_2$ is partitioned into the orthogonal spaces $\mathcal{Y}_{21}$ and $\mathcal{Y}_{22}$. Thus, at this point, $\mathcal{L}_2(\mathrm{IR})$ has been partitioned into the four orthogonal spaces $\mathcal{Y}_{11}$, $\mathcal{Y}_{12}$, $\mathcal{Y}_{21}$, and $\mathcal{Y}_{22}$, where each of these spaces corresponds to the union of certain TF regions R_i.

3. In the next pass, the region S_{11} is partitioned into regions S_{111} and S_{112}, and the associated space $\mathcal{Y}_{11}$ is partitioned into spaces $\mathcal{Y}_{111}$ and $\mathcal{Y}_{112}$. An analogous splitting-up is done for the regions S_{12}, S_{21}, and S_{22} and the associated spaces $\mathcal{Y}_{12}$, $\mathcal{Y}_{21}$, and $\mathcal{Y}_{22}$. This process of successively splitting up super-regions and super-spaces into two new regions or spaces continues until

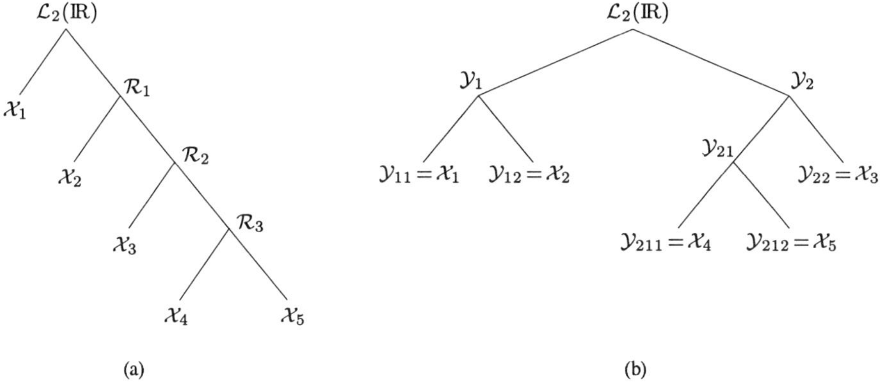

(a)   (b)

Figure 5.8. Tree representation of partitioning strategies—example for $M = 5$: (a) sequential partitioning (cf. Fig. 5.6), (b) binary partitioning (cf. Fig. 5.7).

all super-regions contain only a single region R_i. The spaces corresponding to the elementary TF regions R_i are then the looked-for spaces $\mathcal{X}_i$.

Tree representations of the two partitioning strategies are shown in **Fig. 5.8** for the examples considered in Figs. 5.6 and 5.7. We note that the sequential partitioning method is formally an extreme special case of the binary partitioning method, where at each stage one of the super-regions is chosen as an elementary region R_i. Evidently, many variations of these two partitioning strategies are possible. The construction of such partitioning schemes is equivalent to the coding of a number of elements (the TF regions R_i) by binary codewords of possibly different lengths [Cover and Thomas, 1991]. Irrespectively of the particular partitioning strategy chosen, the core of the design method is the subspace-constrained space synthesis method of Section 4.3.

5.3.4 Simulation Results

Results obtained with a TF projection filter bank corresponding to $M = 3$ TF regions R_i are shown in **Fig. 5.9**. (For $M = 3$, the sequential and binary partitioning schemes are equivalent.) The input signal $x(t)$ is three-component, with all components well concentrated inside one of the TF pass regions R_i. It is seen that the TF projection filter bank successfully separates the three components with only small distortion of the individual components.[9]

---

[9]Due to the perfect-reconstruction property, the sum of all filter outputs equals the input signal even though the individual signal components are distorted in general.

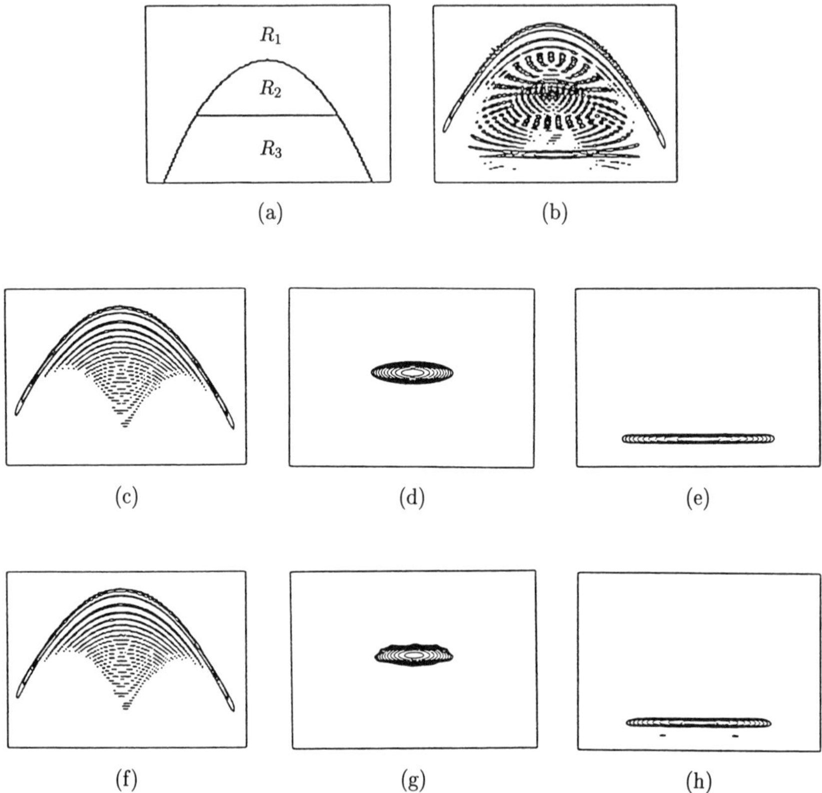

Figure 5.9.   Decomposition of a three-component signal using a TF projection filter bank with $M = 3$: (a) TF partition, (b) WD of three-component input signal, (c)-(e) WDs of individual input signal components, (f)-(h) WDs of individual filter output signals.

In **Fig. 5.10**, the same TF projection filter bank is applied to a one-component chirp signal that extends over all three TF pass regions R_i. The individual filter output signals are effectively confined to the pass regions, with broadening or "ringing" effects (generalizing Gibbs' phenomenon) along the pass regions' boundaries. The sum of all output signals is seen to equal the input signal.

5.4   Discussion

An alternative to the TF projection filter discussed in Sections 5.1 and 5.2 is given by the *Weyl filter* which can be considered as a crude approximation to

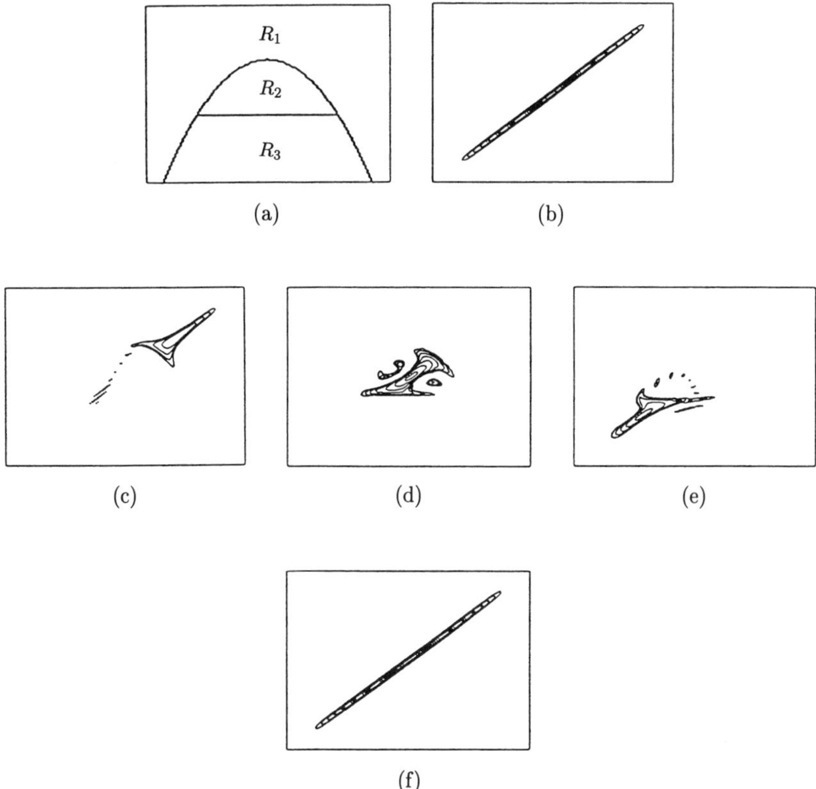

Figure 5.10.   Decomposition of a one-component chirp signal using the TF projection filter bank from Fig. 5.9: (a) TF partition, (b) WD of input signal, (c)-(e) WDs of individual filter output signals, (f) WD of sum of all output signals.

the TF projection filter [Kozek, 1992b, Kozek and Hlawatsch, 1992, Kozek and Hlawatsch, 1991a]. The Weyl filter does not have a projection structure, but its design does not require the solution of an eigenproblem. The Weyl filter is easily extended to perfect-reconstruction TF filter banks [Kozek and Hlawatsch, 1991a]. Other linear TF filters are based on the short-time Fourier transform [Portnoff, 1980, Daubechies, 1988, Bourdier et al., 1988], the Gabor expansion [Farkash and Raz, 1994], the wavelet transform [Daubechies and Paul, 1988], or the WD of a linear system [Hlawatsch, 1992b] (see Section 5.1.3). In this context, we emphasize that the *orthogonal projection structure* of the TF projection filter is appropriate only as long as the filtering task is to pass or reject signals in prescribed TF regions, i.e., to perform a "TF weighting" by weighting factors 1 (pass region) or 0 (stop region). If some more general TF weighting characteristic is desired, then an orthogonal projection structure is no longer adequate and a more general linear filter must be used.

Finally, it should be noted that the TF filters, TF filter banks, and TF-localized basis systems discussed here are practically suited to the *off-line* processing of signals with moderate duration. Indeed, these methods are based on an eigendecomposition of a matrix whose size equals (or is proportional to) the signal length. This obviously entails a practical limitation of the signal length and is incompatible with an on-line mode of operation. However, we just note that for the on-line processing of signals with arbitrary (possibly infinite) duration, a short-time version of the TF projection filter can be developed in which a "local TF space" is continually updated. Other short-time filtering methods are based on the Fourier transform, Gabor expansion, or wavelet transform.

6 SIGNAL ESTIMATION AND SIGNAL DETECTION

The previous chapter has introduced and discussed TF projection filters that pass signals located in a given TF region and suppress signals located outside this TF region. So far, these filters have been studied mainly in a deterministic setting. In this chapter, we shall consider a stochastic setting and discuss applications of TF projection filters in statistical signal processing; here, the signals involved are modeled as (generally nonstationary) random processes.

In Section 6.1, we consider *signal estimation* (enhancement), specifically, the estimation of a nonstationary signal process contaminated by a nonstationary, generally non-white noise process. We derive the optimum linear system for signal estimation under a projection side constraint, i.e., the system is constrained to correspond to an orthogonal projection operator. The solution obtained suggests an intuitively appealing, approximate *TF design* of the optimum projection operator. This TF design amounts to finding the TF projection filter for a TF pass region that is determined by the Wigner-Ville spectra of signal and noise.

Both the optimum projection operator and the approximate TF projection filter can be applied to a seemingly different problem, namely, the estimation of a signal that is known to be located in a given TF region.

Section 6.2 studies *signal detection*, i.e., the detection of a nonstationary signal process contaminated by white Gaussian noise. Again, we consider the use of an orthogonal projection operator. The deflection-optimum projection system is derived and the solution obtained is again seen to suggest an intuitively appealing, approximate TF design. This TF design yields a TF projection filter whose pass region is defined by thresholding the Wigner-Ville spectrum of the signal. Again, these results can be applied to the problem of detecting a signal that is known to be located in a given TF region.

6.1 Signal Estimation

We consider the following estimation problem [Van Trees, 1968, Therrien, 1992, Papoulis, 1984a]: A nonstationary, zero-mean random signal $s(t)$ is contaminated by nonstationary, zero-mean, additive noise $n(t)$ that is uncorrelated with $s(t)$. The autocorrelation functions[1] of $s(t)$ and $n(t)$,

$$R_s(t,t') = \mathrm{E}\{s(t)\,s^*(t')\} \qquad \text{and} \qquad R_n(t,t') = \mathrm{E}\{n(t)\,n^*(t')\}\,,$$

are assumed to be known. Based on the observed (noise-contaminated) process $r(t) = s(t) + n(t)$, we form an estimate $\hat{s}(t)$ of the signal process $s(t)$ using a linear, generally time-varying system $\mathbf{H}$ with impulse response $H(t,t')$,

$$\hat{s}(t) = (\mathbf{H}\,r)(t) = \int_{t'} H(t,t')\,r(t')\,dt'\,.$$

The optimum system minimizing the mean-square error, $\mathrm{E}\{|e(t)|^2\}$ with $e(t) = s(t) - \hat{s}(t)$, for all t is the (nonstationary) *Wiener filter* [Van Trees, 1968, Therrien, 1992]

$$\mathbf{H}_{\mathrm{opt}} = \mathbf{R}_s(\mathbf{R}_s + \mathbf{R}_n)^{-1}\,,$$

where $\mathbf{R}_s$ and $\mathbf{R}_n$ are the autocorrelation operators[2] of signal $s(t)$ and noise $n(t)$, respectively. The calculation of the Wiener filter requires the inversion of the operator $\mathbf{R}_r = \mathbf{R}_s + \mathbf{R}_n$, which may be impossible or numerically unstable.

6.1.1 *Optimum Projection Filter for Signal Estimation*

We now study the solution of the estimation problem with $\mathbf{H}$ constrained to be an orthogonal projection operator. We shall see that this approach avoids the inversion of an operator although it requires the calculation of eigenvalues

---

[1]We assume throughout this chapter that $s(t)$ and $n(t)$ are *circular*, complex-valued processes, i.e., $\mathrm{E}\{s(t)\,s(t')\} \equiv 0$ and $\mathrm{E}\{n(t)\,n(t')\} \equiv 0$, so that their second-order statistics are completely specified by the autocorrelation functions $R_s(t,t')$ and $R_n(t,t')$, respectively.
[2]The autocorrelation operator $\mathbf{R}_x$ of a nonstationary process $x(t)$ is the linear operator whose kernel is the autocorrelation function $R_x(t,t') = \mathrm{E}\{x(t)\,x^*(t')\}$.

and eigenfunctions. We emphasize that the orthogonal projection constraint will generally result in a larger mean-square error. On the other hand, we shall see in Section 6.1.3 that the optimum projection system can be approximated by a very simple and intuitive *TF design* in which the requirements regarding the statistical *a priori* knowledge are relaxed.

Due to our orthogonal projection constraint, the signal estimate is formed as the orthogonal projection of the observed signal $r(t) = s(t) + n(t)$ onto a linear signal space $\mathcal{S}$,

$$\hat{s}(t) = (\mathbf{P}_{\mathcal{S}}\, r)(t)\,,$$

and this space $\mathcal{S}$ is determined such that the resulting mean-square error is minimized for all t,

$$\mathcal{S}_{\mathrm{opt}} = \arg\min_{\mathcal{S}} \mathrm{E}\big\{|e(t)|^2\big\}\,, \qquad e(t) = s(t) - \hat{s}(t)\,.$$

An equivalent optimality criterion is that of minimum mean error energy[3]

$$\overline{E}_e = \int_t \mathrm{E}\big\{|e(t)|^2\big\}\, dt = \int_t R_e(t,t)\, dt = \mathrm{tr}\{\mathbf{R}_e\}$$

(assuming that $\overline{E}_e$ is finite). The optimum space is characterized in the next theorem whose proof is given in Appendix 6.A.

Theorem 6.1 *Let $\mathbf{R}_s$ and $\mathbf{R}_n$ be the known autocorrelation operators of the zero-mean, uncorrelated signal and noise processes, respectively. The space*

$$\mathcal{S}_{\mathrm{opt}} = \arg\min_{\mathcal{S}} \big\{\overline{E}_e\big\}$$

minimizing the mean energy of the estimation error $e(t) = s(t) - \hat{s}(t)$ with $\hat{s}(t) = (\mathbf{P}_{\mathcal{S}}\, r)(t)$ is spanned by the eigenfunctions $u_k^{(D)}(t)$ corresponding to all positive eigenvalues $\lambda_k^{(D)}$ of the linear operator $\mathbf{D} \overset{\triangle}{=} \mathbf{R}_s - \mathbf{R}_n$,

$$\mathcal{S}_{\mathrm{opt}} = \mathrm{span}\big\{u_k^{(D)}(t)\big\}_{k=1}^{N_+}\,,$$

where N_+ denotes the number of positive eigenvalues $\lambda_k^{(D)}$ (these are the N_+ first $\lambda_k^{(D)}$ since the $\lambda_k^{(D)}$ are assumed to be arranged in non-increasing order). The resulting minimum mean error energy is given by

$$\overline{E}_{e,\mathrm{min}} = \overline{E}_e\Big|_{\mathcal{S}=\mathcal{S}_{\mathrm{opt}}} = \overline{E}_s - \sum_{k=1}^{N_+}\lambda_k^{(D)}\,, \tag{6.1}$$

---

[3]The *trace* of a linear operator $\mathbf{H}$ with kernel $H(t,t')$ is defined as $\mathrm{tr}\{\mathbf{H}\} = \int_t H(t,t)\, dt$.

where $\overline{E}_s = \text{tr}\{\mathbf{R}_s\} = \int_t R_s(t,t)\,dt$ is the mean signal energy.

The optimum space $\mathcal{S}_{\text{opt}}$ allows some interesting interpretations:

1. We desire the projection filter $\mathbf{P}_{\mathcal{S}}$ to introduce minimum signal distortion while passing a minimum amount of noise,

$$(\mathbf{P}_{\mathcal{S}}\,s)(t) \approx s(t) \qquad \text{and} \qquad (\mathbf{P}_{\mathcal{S}}\,n)(t) \approx 0.$$

With $\mathbf{P}_{\overline{\mathcal{S}}} = \mathbf{I} - \mathbf{P}_{\mathcal{S}}$ denoting the orthogonal projection operator on the orthogonal complement space $\overline{\mathcal{S}}$, the first relation can be rephrased as $(\mathbf{P}_{\overline{\mathcal{S}}}\,s)(t) \approx 0$. Let us combine both desired properties into the following joint optimality criterion:

$$\mathcal{S}_{\text{opt}} = \arg\min_{\mathcal{S}} \left\{ \text{E}\{\|\mathbf{P}_{\overline{\mathcal{S}}}\,s\|^2\} + \text{E}\{\|\mathbf{P}_{\mathcal{S}}\,n\|^2\} \right\}. \tag{6.2}$$

It is easily shown that the quantity to be minimized can be expressed as

$$\begin{aligned}
\text{E}\{\|\mathbf{P}_{\overline{\mathcal{S}}}\,s\|^2\} + \text{E}\{\|\mathbf{P}_{\mathcal{S}}\,n\|^2\} &= \text{tr}\{\mathbf{P}_{\overline{\mathcal{S}}}\,\mathbf{R}_s\} + \text{tr}\{\mathbf{P}_{\mathcal{S}}\,\mathbf{R}_n\} \\
&= \text{tr}\{\mathbf{R}_s\} - \text{tr}\{\mathbf{P}_{\mathcal{S}}\,\mathbf{R}_s\} + \text{tr}\{\mathbf{P}_{\mathcal{S}}\,\mathbf{R}_n\} \\
&= \text{tr}\{\mathbf{R}_s\} - \text{tr}\{\mathbf{P}_{\mathcal{S}}\,\mathbf{D}\}.
\end{aligned}$$

With $\text{tr}\{\mathbf{R}_s\} = \overline{E}_s$ and $\overline{E}_e = \overline{E}_s - \text{tr}\{\mathbf{P}_{\mathcal{S}}\,\mathbf{D}\}$ according to Eq. (6.A.1) in Appendix 6.A, it follows that $\text{E}\{\|\mathbf{P}_{\overline{\mathcal{S}}}\,s\|^2\} + \text{E}\{\|\mathbf{P}_{\mathcal{S}}\,n\|^2\} = \overline{E}_e$. Hence, the optimality criterion (6.2) is identical to that of minimizing the mean error energy $\overline{E}_e$, and thus the optimum space defined by (6.2) equals the optimum space considered in Theorem 6.1.

2. The quantities $\text{E}\{\|\mathbf{P}_{\mathcal{S}_{\text{opt}}}\,s\|^2\}$ and $\text{E}\{\|\mathbf{P}_{\mathcal{S}_{\text{opt}}}\,n\|^2\}$ can be interpreted as the mean energies of those signal and noise components, respectively, that fall into the space $\mathcal{S}_{\text{opt}}$. It is easily shown that

$$\text{E}\{\|\mathbf{P}_{\mathcal{S}_{\text{opt}}}\,s\|^2\} = \sum_{k=1}^{N_+} \text{E}\{|\langle s, u_k^{(D)} \rangle|^2\} \tag{6.3}$$

$$\text{E}\{\|\mathbf{P}_{\mathcal{S}_{\text{opt}}}\,n\|^2\} = \sum_{k=1}^{N_+} \text{E}\{|\langle n, u_k^{(D)} \rangle|^2\}. \tag{6.4}$$

Here, $\text{E}\{|\langle s, u_k^{(D)} \rangle|^2\}$ and $\text{E}\{|\langle n, u_k^{(D)} \rangle|^2\}$ can be interpreted as the mean signal and noise energy, respectively, falling into the direction of the kth eigenfunction $u_k^{(D)}(t)$. Now there is

$$\begin{aligned}
\mathrm{E}\{|\langle s, u_k^{(D)}\rangle|^2\} - \mathrm{E}\{|\langle n, u_k^{(D)}\rangle|^2\} &= \langle \mathbf{R}_s\, u_k^{(D)}, u_k^{(D)}\rangle - \langle \mathbf{R}_n\, u_k^{(D)}, u_k^{(D)}\rangle \\
&= \langle \mathbf{D}\, u_k^{(D)}, u_k^{(D)}\rangle = \lambda_k^{(D)} \\
&> 0 \quad \text{for} \quad k = 1, ..., N_+
\end{aligned}$$

or equivalently

$$\mathrm{E}\{|\langle s, u_k^{(D)}\rangle|^2\} > \mathrm{E}\{|\langle n, u_k^{(D)}\rangle|^2\} \quad \text{for} \quad k = 1, ..., N_+ \,.$$

Hence, in the directions of the eigenfunctions $u_k^{(D)}(t)$ ($k = 1, ..., N_+$) spanning $\mathcal{S}_{\mathrm{opt}}$, the mean signal energy is larger than the mean noise energy. In other words, the optimum space is spanned by all eigenfunction directions in which the mean signal energy is larger than the mean noise energy. With (6.3) and (6.4), we also obtain

$$\mathrm{E}\{\|\mathbf{P}_{\mathcal{S}_{\mathrm{opt}}} s\|^2\} > \mathrm{E}\{\|\mathbf{P}_{\mathcal{S}_{\mathrm{opt}}} n\|^2\} \,,$$

which shows that within the optimum space $\mathcal{S}_{\mathrm{opt}}$, the mean signal energy is larger than the mean noise energy.

3. In the important special case where the noise $n(t)$ is stationary and white with power spectral density $S_n(f) \equiv \eta$, the autocorrelation operator of $n(t)$ is $\mathbf{R}_n = \eta\,\mathbf{I}$. Let $\lambda_k^{(s)}$ and $u_k^{(s)}(t)$ denote the eigenvalues and eigenfunctions, respectively, of the signal's autocorrelation operator $\mathbf{R}_s$, with the eigenfunctions $u_k^{(s)}(t)$ suitably augmented (if necessary) such that they form an orthonormal basis of $\mathcal{L}_2(\mathbb{R})$. Then, the eigenfunctions of $\mathbf{R}_n = \eta\,\mathbf{I}$ can be chosen to be equal to the signal eigenfunctions $u_k^{(s)}(t)$, with constant noise eigenvalue spectrum $\lambda_k^{(n)} \equiv \eta$ [Van Trees, 1968, Therrien, 1992]. It follows that the eigenfunctions of $\mathbf{D}$ are the signal eigenfunctions, $u_k^{(D)}(t) = u_k^{(s)}(t)$, and the eigenvalues of $\mathbf{D} = \mathbf{R}_s - \mathbf{R}_n$ become[4]

$$\lambda_k^{(D)} = \lambda_k^{(s)} - \lambda_k^{(n)} = \lambda_k^{(s)} - \eta \,.$$

The condition $\lambda_k^{(D)} > 0$ then implies

$$\lambda_k^{(s)} > \eta \quad \text{for} \quad k = 1, ..., N_+ \,,$$

---

[4]We emphasize that the relation $\lambda_k^{(D)} = \lambda_k^{(s)} - \lambda_k^{(n)}$ is valid only if the eigenfunctions of signal and noise are equal.

which means that the mean signal energy $E\{|\langle s, u_k^{(D)}\rangle|^2\} = \lambda_k^{(s)}$ in the direction of any eigenfunction spanning the optimum space $\mathcal{S}_{\text{opt}}$ is larger than the mean noise energy in this direction, η. Equivalently, we can say that the optimum space is spanned by all signal eigenfunctions in whose direction the mean signal energy lies above the noise level η.

6.1.2 Two Special Cases

Further insights can be obtained by considering two special situations in which the optimum projection system is particularly simple and intuitively appealing.

Stationary processes. If $s(t)$ and $n(t)$ are jointly wide-sense stationary with power spectral densities $S_s(f)$ and $S_n(f)$, respectively, then it can be shown that the optimum projection system $\mathbf{P}_{\mathcal{S}_{\text{opt}}}$ is a time-*in*variant filter with binary-valued frequency response

$$H(f) = \begin{cases} 1 & \text{for } f\colon\ S_s(f) > S_n(f) \\ 0 & \text{for } f\colon\ S_s(f) \le S_n(f) . \end{cases}$$

Thus, $\mathbf{P}_{\mathcal{S}_{\text{opt}}}$ passes all frequencies where the signal is stronger than the noise and suppresses all other frequencies (cf. [Cimini and Kassam, 1981]). The signal estimate $\hat{s}(t)$ and the estimation error $e(t) = s(t) - \hat{s}(t)$ are wide-sense stationary processes. The power spectral density of the estimation error can be shown to be

$$\begin{aligned} S_e(f) &= S_s(f) - H(f)\left[S_s(f) - S_n(f)\right] \\[2mm] &= \begin{cases} S_n(f) & \text{for } f\colon\ S_s(f) > S_n(f) \\ S_s(f) & \text{for } f\colon\ S_s(f) < S_n(f) \text{ or } S_s(f) = S_n(f) \ne 0 \\ 0 & \text{for } f\colon\ S_s(f) = S_n(f) = 0 . \end{cases} \end{aligned}$$

These results are consistent with the general results in Theorem 6.1. Indeed, stationary processes can be considered a limiting case where

$$\begin{aligned} k &\longrightarrow f \\ \lambda_k^{(s)} &\longrightarrow S_s(f) \\ \lambda_k^{(n)} &\longrightarrow S_n(f) \\ \lambda_k^{(D)} &\longrightarrow S_s(f) - S_n(f) \\ u_k^{(s)}(t),\ u_k^{(n)}(t),\ u_k^{(D)}(t) &\longrightarrow e^{j2\pi f t} \end{aligned}$$

$$P_{\mathcal{S}_{\mathrm{opt}}}(t_1, t_2) = \sum_{k:\,\lambda_k>0} u_k^{(D)}(t_1)\, u_k^{(D)*}(t_2) \quad \longrightarrow \quad \int_{f:\,S_s(f)-S_n(f)>0} e^{j2\pi f t_1}\, e^{-j2\pi f t_2}\, df$$

$$= \int_f H(f)\, e^{j2\pi f(t_1-t_2)}\, df.$$

Note, however, that the results of Theorem 6.1 are not directly applicable since the mean error energy $\overline{E}_e$ is not finite.

Nonstationary white processes. An analogous situation is that of nonstationary white $s(t)$ and $n(t)$ with average instantaneous intensities[5] $q_s(t)$ and $q_n(t)$, respectively. Here, the optimum projection system $\mathbf{P}_{\mathcal{S}_{\mathrm{opt}}}$ is a "frequency-invariant" system corresponding to a time gating,

$$\hat{s}(t) = \big(\mathbf{P}_{\mathcal{S}_{\mathrm{opt}}} r\big)(t) = h(t)\, r(t) \qquad \text{with} \quad h(t) = \begin{cases} 1 & \text{for } t:\ q_s(t) > q_n(t) \\ 0 & \text{for } t:\ q_s(f) \le q_n(f). \end{cases}$$

Thus, the optimum projection filter passes all time intervals where the signal is stronger than the noise and suppresses all other times. The estimation error $e(t) = s(t) - \hat{s}(t)$ is nonstationary white with average instantaneous intensity

$$\begin{aligned} q_e(t) &= q_s(t) - h(t)\,[q_s(t) - q_n(t)] \\ &= \begin{cases} q_n(t) & \text{for } t:\ q_s(t) > q_n(t) \\ q_s(t) & \text{for } t:\ q_s(t) < q_n(t) \ \text{or}\ q_s(t) = q_n(t) \neq 0 \\ 0 & \text{for } t:\ q_s(t) = q_n(t) = 0. \end{cases} \end{aligned}$$

Again, these results are consistent with the general results in Theorem 6.1, with

$$k \ \longrightarrow\ t$$
$$\lambda_k^{(s)} \ \longrightarrow\ q_s(t)$$
$$\lambda_k^{(n)} \ \longrightarrow\ q_n(t)$$
$$\lambda_k^{(D)} \ \longrightarrow\ q_s(t) - q_n(t)$$
$$u_k^{(s)}(t'),\, u_k^{(n)}(t'),\, u_k^{(D)}(t') \ \longrightarrow\ \delta(t'-t)$$
$$P_{\mathcal{S}_{\mathrm{opt}}}(t_1, t_2) = \sum_{k:\,\lambda_k>0} u_k^{(D)}(t_1)\, u_k^{(D)*}(t_2) \ \longrightarrow\ \int_{t:\,q_s(t)-q_n(t)>0} \delta(t_1-t)\,\delta(t_2-t)\,dt$$
$$= h(t)\,\delta(t_1-t_2).$$

---

[5]The autocorrelation function of a nonstationary white process $x(t)$ is $R_x(t,t') = q_x(t)\,\delta(t-t')$ where $q_x(t) \ge 0$ is called the average instantaneous intensity of $x(t)$ [Papoulis, 1984a].

6.1.3   Time-Frequency Design

The special cases studied in the last subsection suggest a heuristic *TF design* of projection filters for signal estimation. Recall that, for stationary processes, the optimum projection filter passes all frequencies f where $S_s(f) > S_n(f)$. Similarly, for nonstationary white processes, the optimum projection filter passes all times t where $q_s(t) > q_n(t)$. In the general case of nonstationary processes, we may attempt to extend the pure frequency localization or pure time localization of these two special situations to a joint TF localization. Specifically, we look for a projection filter that passes all TF points (t, f) where the signal is stronger than the noise and suppresses all other TF points. This requires a means for characterizing the mean TF energy distribution of a nonstationary process. For this purpose we use the *Wigner-Ville spectrum* (WVS) [Martin and Flandrin, 1985, Flandrin, 1989, Flandrin and Martin, 1997]. The WVS of a nonstationary process $x(t)$ with autocorrelation function $R_x(t, t')$ is formally defined as the expectation of the process' WD (cf. Section 2.1.2),

$$\overline{W}_x(t, f) \triangleq \mathrm{E}\{W_x(t, f)\} = \int_\tau R_x\left(t + \frac{\tau}{2}, t - \frac{\tau}{2}\right) e^{-j2\pi f\tau}\, d\tau = L_{\mathbf{R}_x}(t, f)\,,$$

which is recognized as the Weyl symbol of the autocorrelation operator $\mathbf{R}_x$. The WVS $\overline{W}_x(t, f)$ reduces to the power spectral density $S_x(f)$ in the case of a stationary process and to the average instantaneous intensity $q_x(t)$ in the case of a nonstationary white process.

We now consider the TF region on which the signal's WVS is larger than the WVS of the noise,

$$R_D \triangleq \left\{(t, f): \overline{W}_s(t, f) > \overline{W}_n(t, f)\right\}.$$

Then, the TF projection filter corresponding to R_D in the sense of Chapter 5 may be expected to perform a satisfactory estimation of the signal process $s(t)$, and hence to be a reasonable approximation to the optimum projection filter described in Section 6.1.1. We recall that this TF projection filter is the orthogonal projection operator $\mathbf{P}_{\mathcal{U}_{R_D}}$ on the eigenspace $\mathcal{U}_{R_D}$ of R_D, i.e., the space whose WD is closest to the indicator function of R_D given by

$$I_{R_D}(t, f) \triangleq \begin{cases} 1 & \text{for } (t, f): \overline{W}_s(t, f) > \overline{W}_n(t, f) \\ 0 & \text{for } (t, f): \overline{W}_s(t, f) \le \overline{W}_n(t, f)\,. \end{cases}$$

However, the heuristic TF projection filter defined above will not always be a good approximation to the optimum projection filter. In order to investigate this issue, we first establish an interpretation of the TF pass region R_D in terms of the difference operator $\mathbf{D} = \mathbf{R}_s - \mathbf{R}_n$. The Weyl symbol of $\mathbf{D}$,

$$L_{\mathbf{D}}(t, f) = \int_\tau D\left(t + \frac{\tau}{2}, t - \frac{\tau}{2}\right) e^{-j2\pi f\tau} \, d\tau \, ,$$

is given by the difference of the WVS of signal and noise,

$$L_{\mathbf{D}}(t, f) = L_{\mathbf{R}_s}(t, f) - L_{\mathbf{R}_n}(t, f) = \overline{W}_s(t, f) - \overline{W}_n(t, f) \, .$$

Hence, the condition $\overline{W}_s(t, f) > \overline{W}_n(t, f)$ defining the TF pass region R_D can be reformulated as $L_{\mathbf{D}}(t, f) > 0$, i.e.,

$$R_D = \left\{ (t, f) : \; L_{\mathbf{D}}(t, f) > 0 \right\} .$$

Thus, the TF projection filter passes all TF points for which $L_{\mathbf{D}}(t, f)$ is positive.

We now recall that the *optimum* projection filter performs a projection onto the space $\mathcal{S}_{\mathrm{opt}}$ spanned by all eigenfunctions $u_k^{(D)}(t)$ of the operator $\mathbf{D}$ whose corresponding eigenvalues are positive, $\lambda_k^{(D)} > 0$. Does there exist a relation between the Weyl symbol of $\mathbf{D}$ (defining the TF projection filter) and the eigenvalues and eigenfunctions of $\mathbf{D}$ (defining the optimum projection filter)? Indeed, it is well known that

$$\int_t \int_f L_{\mathbf{D}}(t, f) \, W_{u_k^{(D)}}(t, f) \, dt \, df = \langle \mathbf{D} u_k^{(D)}, u_k^{(D)} \rangle = \lambda_k^{(D)} \, .$$

If $u_k^{(D)}(t)$ is well TF concentrated in a TF region R_k, then this relation can be interpreted in the sense that a local average of $L_{\mathbf{D}}(t, f)$ over the TF region R_k equals $\lambda_k^{(D)}$. If $L_{\mathbf{D}}(t, f)$ is a sufficiently smooth function, this implies a local approximate equality of $L_{\mathbf{D}}(t, f)$ and $\lambda_k^{(D)}$:

$$L_{\mathbf{D}}(t, f) \approx \lambda_k^{(D)} \qquad \text{for } (t, f) \in R_k \, .$$

Hence, including an eigenfunction $u_k^{(D)}(t)$ with $\lambda_k^{(D)} > 0$ in the basis spanning the optimum space $\mathcal{S}_{\mathrm{opt}}$ is approximately equivalent to passing the TF region R_k in which the Weyl symbol of $\mathbf{D}$ is positive, $L_{\mathbf{D}}(t, f) \approx \lambda_k^{(D)} > 0$. This shows that the TF region $R_D = \left\{ (t, f) : \; L_{\mathbf{D}}(t, f) > 0 \right\}$ corresponds approximately to the optimum space $\mathcal{S}_{\mathrm{opt}}$ spanned by all $u_k^{(D)}(t)$ with $\lambda_k^{(D)} > 0$, and thus the TF projection filter is a good approximation to the optimum projection filter.

This discussion shows that a sufficient condition for the approximate equivalence of the TF projection filter and the optimum projection filter is that the Weyl symbol of $\mathbf{D}$ is a smooth TF function, and that the eigenfunctions of $\mathbf{D}$ are well TF concentrated. Both conditions will be met if $\mathbf{D}$ is an *underspread* operator [Kozek, 1997b, Kozek, 1997a]. With $\mathbf{D} = \mathbf{R}_s - \mathbf{R}_n$, this implies in turn that the signal process $s(t)$ and the noise process $n(t)$ are *jointly underspread processes*[6] [Kozek et al., 1994, Kozek, 1996, Kozek, 1998, Kozek, 1997a].

The TF projection filter has two advantages over the optimum projection filter:

- The statistical *a priori* knowledge required for the design of the TF projection filter is formulated directly in the intuitively more accessible TF domain.

- Furthermore, this *a priori* knowledge is reduced since it suffices to know the TF region in which the WVS of the signal is larger than the WVS of the noise; complete knowledge of the two WVS is not required. In contrast, the design of the optimum projection filter presupposes complete knowledge of the autocorrelation operators of signal and noise (this is equivalent to complete knowledge of the WVS of signal and noise).

6.1.4   Simulation Results

We now provide an experimental comparison of the optimum projection filter and the TF projection filter. **Fig. 6.1** shows the WVS of signal $s(t)$ and noise $n(t)$, as well as the WDs of the space $\mathcal{S}_{\mathrm{opt}}$ underlying the optimum projection filter and the space $\mathcal{U}_{R_D}$ underlying the TF projection filter. The WVS of signal and noise are partly overlapping in the TF plane. The WDs of the spaces $\mathcal{S}_{\mathrm{opt}}$ and $\mathcal{U}_{R_D}$ are seen to be very similar, which shows the approximate equivalence of the two projection filters.[7] The mean SNR improvements achieved by the optimum projection filter and the TF projection filter are 3.59 dB and 3.50 dB, respectively. Hence, it is seen that the TF projection filter performs nearly as well as the optimum projection filter. The comparatively small SNR improvements are due to the fact that signal and noise overlap significantly in the TF plane. Realizations of all signals involved are shown in **Fig. 6.2**. We note that the processes $s(t)$ and $n(t)$ were synthesized using the TF synthesis method proposed in [Hlawatsch and Kozek, 1995].

---

[6]A nonstationary process is underspread [Kozek et al., 1994, Kozek, 1996, Kozek, 1998, Kozek, 1997a] if process components that are sufficiently distant in the TF plane are effectively uncorrelated. This can be shown to entail a smooth WVS. Note that smoothness of the WVS $\overline{W}_s(t, f)$ and $\overline{W}_n(t, f)$ entails smoothness of $L_{\mathbf{D}}(t, f) = \overline{W}_s(t, f) - \overline{W}_n(t, f)$.

[7]For numerical reasons, the condition $\lambda_k^{(D)} > 0$ defining the optimum projection filter and the condition $L_{\mathbf{D}}(t, f) > 0$ defining the TF projection filter have been replaced by the conditions $\lambda_k^{(D)} > \max_k\{\lambda_k^{(D)}\}/100$ and $L_{\mathbf{D}}(t, f) > \max_{t,f}\{L_{\mathbf{D}}(t, f)\}/100$, respectively. This explains why both filters suppress those TF regions where the WVS of signal and noise have approximately equal heights, i.e., in particular, where both WVS are zero and also (in the special case considered) where both WVS overlap.

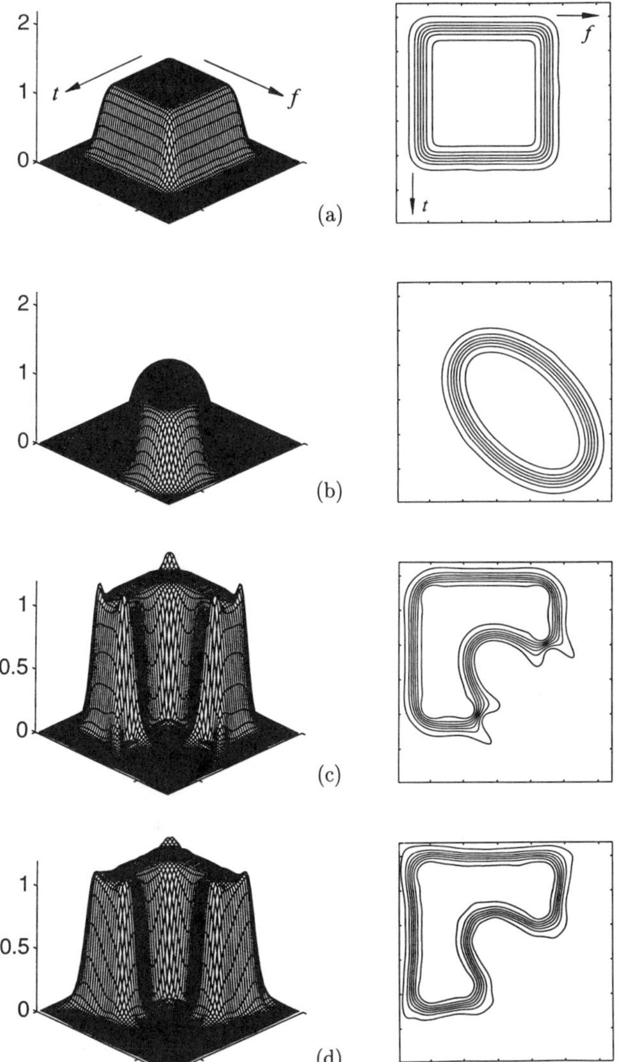

Figure 6.1. Comparison of optimum projection filter and TF projection filter for signal estimation (I): (a) WVS of signal $s(t)$, (b) WVS of noise $n(t)$, (c) WD of space $\mathcal{S}_{\mathrm{opt}}$ underlying the optimum projection filter, and (d) WD of space $\mathcal{U}_{R_D}$ underlying the TF projection filter. We note that a slight smoothing has been applied to all WVS and WDs.

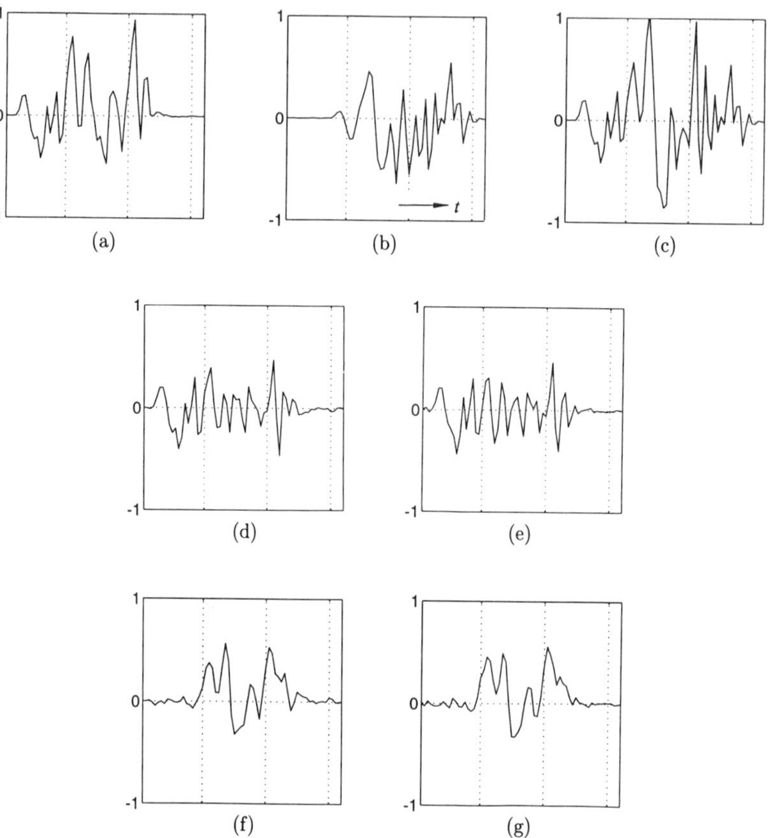

Figure 6.2. Comparison of optimum projection filter and TF projection filter for signal estimation (II): Real parts of realizations of (a) signal $s(t)$, (b) noise $n(t)$, (c) observed (noisy) signal $r(t) = s(t) + n(t)$, (d) signal estimate $\hat{s}(t)$ of optimum projection filter, (e) signal estimate $\hat{s}(t)$ of TF projection filter, (f) error signal $e(t) = s(t) - \hat{s}(t)$ of optimum projection filter, (g) error signal $e(t) = s(t) - \hat{s}(t)$ of TF projection filter.

6.1.5 Estimation of Time-Frequency Localized Signals

The results developed above can be applied to a different estimation problem. Let us consider the case where no statistical *a priori* knowledge about the signal $s(t)$ (such as its autocorrelation function $R_s(t, t')$) is available, but it is known that $s(t)$ is located within a given TF region R, and furthermore the energy

E_s of $s(t)$ is known. As before, the additive noise $n(t)$ is characterized by its autocorrelation function $R_n(t,t')$. The problem is again to find an orthogonal projection filter for estimating $s(t)$ given $r(t) = s(t) + n(t)$.

A simple solution to this problem is the TF projection filter associated to the TF region R in which the signal $s(t)$ is known to be located (see Section 5.2). However, this solution completely ignores the available *a priori* knowledge about the noise. It is clear that the TF projection filter associated to R will not perform well if a large part of the noise falls into some subspace of the eigenspace $\mathcal{U}_R$. Therefore, we shall look for a better solution.

We first develop a stochastic model for the signal $s(t)$. It is reasonable to model the known TF localization of $s(t)$ in R by assuming that $s(t) \in \mathcal{U}_R$ or

$$s(t) = \sum_{k=1}^{N_R} s_k \, u_k(t) \,, \tag{6.5}$$

where $\{u_k(t)\}_{k=1}^{N_R}$ is the eigensignal basis spanning $\mathcal{U}_R$. In the absence of additional information about the signal, it is furthermore reasonable to model the coefficients s_k as being statistically independent, zero-mean random variables with identical average powers $\mathrm{E}\{|s_k|^2\} = \lambda^{(s)}$, so that

$$\mathrm{E}\{s_k s_l^*\} = \lambda^{(s)} \delta_{kl} \,. \tag{6.6}$$

Note that the average powers must be identical since otherwise certain directions of our eigenspace would be assigned more signal power than other directions, which is not supported by the *a priori* knowledge available. With (6.5) and (6.6), the autocorrelation function of $s(t)$ is obtained as

$$R_s(t,t') = \lambda^{(s)} \sum_{k=1}^{N_R} u_k(t) \, u_k^*(t') = \lambda^{(s)} \, P_{\mathcal{U}_R}(t,t') \,.$$

Up to a factor $\lambda^{(s)}$, this is the kernel of the orthogonal projection operator on $\mathcal{U}_R$. Setting the average energy of our stochastic signal model, $\overline{E}_s = \int_t R_s(t,t) \, dt = \lambda^{(s)} N_R$, equal to the known signal energy E_s, we obtain $\lambda^{(s)} = E_s/N_R$ and finally

$$\mathbf{R}_s = \frac{E_s}{N_R} \, \mathbf{P}_{\mathcal{U}_R} \,. \tag{6.7}$$

The second-order statistics of our signal model are now completely specified, and it is thus possible to apply the results of Theorem 6.1. Hence, the optimum space onto which to project $r(t)$ is spanned by the eigenfunctions $u_k^{(D)}(t)$ corresponding to all positive eigenvalues $\lambda_k^{(D)}$ of the operator

$$\mathbf{D} = \mathbf{R}_s - \mathbf{R}_n = \frac{E_s}{N_R} \, \mathbf{P}_{\mathcal{U}_R} - \mathbf{R}_n \,.$$

The resulting projection operator takes into account the *a priori* knowledge about the noise (as expressed by $\mathbf{R}_n$) and, in general, it is of course different from the naive solution $\mathbf{P}_{\mathcal{U}_R}$. However, in the special case that the noise is stationary white, $\mathbf{R}_n = \eta \mathbf{I}$, and the signal eigenvalues are larger than the noise eigenvalues, $E_s/N_R > \eta$, we reobtain the naive solution $\mathbf{P}_{\mathcal{U}_R}$.

An approximate TF design of the estimator derived above is easily obtained. According to Section 6.1.3, we can approximate the optimum projection operator by the TF projection filter corresponding to the TF region $R_D \triangleq \{(t,f) : \overline{W}_s(t,f) > \overline{W}_n(t,f)\}$. With (6.7), we obtain

$$\overline{W}_s(t,f) = \frac{E_s}{N_R} L_{\mathbf{P}_{\mathcal{U}_R}}(t,f) = \frac{E_s}{N_R} W_{\mathcal{U}_R}(t,f),$$

where $W_{\mathcal{U}_R}(t,f)$ is the WD of the eigenspace $\mathcal{U}_R$. Hence, the region R_D becomes

$$R_D = \left\{(t,f) : \frac{E_s}{N_R} W_{\mathcal{U}_R}(t,f) > \overline{W}_n(t,f)\right\}.$$

Since $W_{\mathcal{U}_R}(t,f)$ is similar to the indicator function $I_R(t,f)$ apart from small-scale oscillations, it is reasonable to replace R_D by the simplified region

$$\tilde{R} \triangleq \left\{(t,f) : \frac{E_s}{N_R} I_R(t,f) > \overline{W}_n(t,f)\right\}.$$

Assuming $\overline{W}_n(t,f)$ to be nonnegative, this region may not contain TF points outside R; hence, it consists of all TF points (t,f) within R where the average signal energy density, E_s/N_R, is larger than the average noise energy density:

$$\tilde{R} = \left\{(t,f) \in R : \frac{E_s}{N_R} > \overline{W}_n(t,f)\right\},$$

which is an intuitively pleasing result.

6.2 Signal Detection

We next consider the problem of detecting the presence of a random signal $s(t)$ in an observed process $r(t)$. The statistical hypotheses are

$$\begin{aligned} H_0 : \quad r(t) &= n(t) \\ H_1 : \quad r(t) &= s(t) + n(t). \end{aligned}$$

Let us temporarily assume that signal $s(t)$ and noise $n(t)$ are statistically independent, nonstationary, zero-mean, circularly symmetric processes with known autocorrelation functions $R_s(t,t')$ and $R_n(t,t')$, respectively. In addition, we also assume that $s(t)$ and $n(t)$ are complex *Gaussian* processes. Then, it is well known [Van Trees, 1992, Whalen, 1971, Poor, 1988] that the likelihood ratio detector (which optimizes various performance measures based on the probabilities of detection and false alarm) forms the detection statistic

$$Q_r^{(\text{opt})} = \langle \mathbf{H}_{\text{opt}}\, r,\, r \rangle$$

with[8]

$$\mathbf{H}_{\text{opt}} = \mathbf{R}_n^{-1}\mathbf{R}_s(\mathbf{R}_s + \mathbf{R}_n)^{-1} = \mathbf{R}_n^{-1} - (\mathbf{R}_s + \mathbf{R}_n)^{-1},$$

and then decides for H_0 if $Q_r^{(\text{opt})} \leq \gamma$ and for H_1 if $Q_r^{(\text{opt})} > \gamma$, where γ is a fixed threshold.

An alternative optimality criterion is that of maximum deflection (or "SNR") [Poor, 1988, Picinbono and Duvaut, 1988, Baker, 1966]

$$d = \frac{\left(\mathrm{E}_1\{Q_r\} - \mathrm{E}_0\{Q_r\}\right)^2}{\mathrm{var}_0\{Q_r\}},$$

where $\mathrm{E}_i\{\cdot\}$ and $\mathrm{var}_i\{\cdot\}$ denote the conditional expectation and conditional variance, respectively, given hypothesis H_i. Assuming a quadratic detection statistic, $Q_r = \langle \mathbf{H}r, r \rangle$, the maximization of d yields[9] [Poor, 1988, Picinbono and Duvaut, 1988, Baker, 1966]

$$Q_r^{(d)} = \langle \mathbf{H}_d\, r,\, r \rangle \qquad \text{with} \quad \mathbf{H}_d = \mathbf{R}_n^{-1}\mathbf{R}_s\,\mathbf{R}_n^{-1}\,.$$

In both cases, the detection statistic is a quadratic form of the observed signal $r(t)$ generated by a positive definite or semi-definite linear operator $\mathbf{H}_{\text{opt}}$ or $\mathbf{H}_d$. With the positive (semi-) definite operator square-root $\mathbf{H}_{\text{opt}}^{1/2}$, the detection statistic $Q_r^{(\text{opt})}$ can also be written as

$$Q_r^{(\text{opt})} = \langle \mathbf{H}_{\text{opt}}\, r,\, r \rangle = \langle \mathbf{H}_{\text{opt}}^{1/2}\, r,\, \mathbf{H}_{\text{opt}}^{1/2}\, r \rangle = \left\| \mathbf{H}_{\text{opt}}^{1/2}\, r \right\|^2,$$

i.e., $Q_r^{(\text{opt})}$ can be obtained by passing $r(t)$ through the system $\mathbf{H}_{\text{opt}}^{1/2}$ and then computing the energy of the output signal $\left(\mathbf{H}_{\text{opt}}^{1/2}\, r\right)(t)$. An analogous interpretation can be given for $Q_r^{(d)}$.

6.2.1 Optimum Projection Filter for Signal Detection

The inversion of the noise autocorrelation operator $\mathbf{R}_n$ required for the calculation of the detection statistic $Q_r^{(\text{opt})}$ or $Q_r^{(d)}$ may cause numerical problems.

---

[8]The inverse of the noise autocorrelation operator $\mathbf{R}_n$ is assumed to exist.

[9]This quadratic detection statistic maximizes the deflection even if the signal process $s(t)$ is not Gaussian.

Therefore, we shall now study the case where the optimum linear operator $\mathbf{H}_{\mathrm{opt}}^{1/2}$ or $\mathbf{H}_d^{1/2}$ is replaced by an orthogonal projection operator $\mathbf{P}_\mathcal{S}$ on some suitably chosen linear signal space $\mathcal{S}$. Similar to the estimation problem studied in Section 6.1, we expect that the projection must be chosen such that it passes a maximum part of the signal and rejects a maximum part of the noise.

With $\mathbf{H} = \mathbf{P}_\mathcal{S}$, the detection statistic is now

$$\tilde{Q}_r = \left\| \mathbf{P}_\mathcal{S}\, r \right\|^2 = \langle \mathbf{P}_\mathcal{S}\, r, r \rangle ,$$

where the last identity is due to the idempotency and self-adjointness of $\mathbf{P}_\mathcal{S}$. The optimization of probabilistic performance measures (involving the probabilities of detection and false alarm) with respect to the signal space $\mathcal{S}$ seems to be difficult. Similarly, we have not been able to maximize the deflection,

$$\tilde{d} = \frac{\left(\mathrm{E}_1\{\tilde{Q}_r\} - \mathrm{E}_0\{\tilde{Q}_r\} \right)^2}{\mathrm{var}_0\{\tilde{Q}_r\}} \qquad \text{with} \quad \tilde{Q}_r = \left\| \mathbf{P}_\mathcal{S}\, r \right\|^2, \tag{6.8}$$

in the general case of nonstationary, nonwhite noise. Therefore, we shall consider the maximization of $\tilde{d}$ only for stationary white noise.[10] Here, the optimum space is characterized in the next theorem whose proof is given in Appendix 6.B. Note that we no longer assume $s(t)$ to be a Gaussian process.

Theorem 6.2 *Let $s(t)$ be a zero-mean, generally nonstationary signal process with known autocorrelation operator $\mathbf{R}_s$, and let $n(t)$ be stationary white Gaussian noise that is uncorrelated with $s(t)$. The space maximizing the deflection,*

$$\mathcal{S}_{\mathrm{opt}} = \arg\max_{\mathcal{S}} \tilde{d} = \arg\max_{\mathcal{S}} \frac{\left(\mathrm{E}_1\{\tilde{Q}_r\} - \mathrm{E}_0\{\tilde{Q}_r\} \right)^2}{\mathrm{var}_0\{\tilde{Q}_r\}} \qquad \text{with} \quad \tilde{Q}_r = \left\| \mathbf{P}_\mathcal{S}\, r \right\|^2,$$

is spanned by the eigenfunctions $u_k^{(s)}(t)$ corresponding to the N_{opt} largest eigenvalues $\lambda_k^{(s)}$ of the signal autocorrelation operator $\mathbf{R}_s$ (these are the N_{opt} first $\lambda_k^{(s)}$ since the $\lambda_k^{(s)}$ are assumed to be arranged in non-increasing order),

$$\mathcal{S}_{\mathrm{opt}} = \mathrm{span}\{u_k^{(s)}(t)\}_{k=1}^{N_{\mathrm{opt}}} .$$

Here, N_{opt} is given by

$$N_{\mathrm{opt}} = \arg\max_N \left\{ \frac{1}{N} \left(\sum_{k=1}^{N} \lambda_k^{(s)} \right)^2 \right\} . \tag{6.9}$$

---

[10] In the general case of nonstationary, nonwhite noise, we may use a whitening filter $\mathbf{R}_n^{-1/2}$ in front of $\mathbf{P}_\mathcal{S}$ to make the noise stationary and white. However, the overall system, $\mathbf{P}_\mathcal{S}\mathbf{R}_n^{-1/2}$, will then no longer be a projection system, and the inversion of $\mathbf{R}_n$ is not avoided.

The maximum deflection achieved is

$$\tilde{d}_{\max} = \tilde{d}\Big|_{\mathcal{S}=\mathcal{S}_{\mathrm{opt}}} = \frac{1}{\eta^2 N_{\mathrm{opt}}} \left(\sum_{k=1}^{N_{\mathrm{opt}}} \lambda_k^{(s)} \right)^2 , \qquad (6.10)$$

where η is the power spectral density of the white noise.

This result is intuitively reasonable, since the space $\mathcal{S}_{\mathrm{opt}}$ is spanned by the signal eigenfunctions corresponding to a set of *largest* signal eigenvalues $\lambda_k^{(s)}$, and therefore a maximum part of the signal is passed. Indeed, as shown in Appendix 6.B, for the optimum space the numerator of $\tilde{d}$ is given by

$$\left(\mathrm{E}_1\{\tilde{Q}_r\} - \mathrm{E}_0\{\tilde{Q}_r\} \right)^2 \Big|_{\mathcal{S}=\mathcal{S}_{\mathrm{opt}}} = \left(\sum_{k=1}^{N_{\mathrm{opt}}} \lambda_k^{(s)} \right)^2 .$$

Since the noise is white, it has equal power in each direction of the total space $\mathcal{L}_2(\mathbb{R})$, and thus it does not influence $\mathcal{S}_{\mathrm{opt}}$. Note that a large dimension of $\mathcal{S}_{\mathrm{opt}}$ is at a penalty since, as shown in Appendix 6.B, the denominator of $\tilde{d}$ (i.e., the variance of $\tilde{Q}_r$ under hypothesis H_0) is proportional to $N_{\mathcal{S}}$, $\mathrm{var}_0\{\tilde{Q}_r\} = \eta^2 N_{\mathcal{S}}$. The optimum space dimension $N_{\mathcal{S}} = N_{\mathrm{opt}}$ in (6.9) is hence an optimum compromise between passing a large part of the signal (by choosing a large dimension) and passing a small part of the noise (by choosing a small dimension). Here, it is interesting that the noise level η does not influence N_{opt}, although of course it affects the maximum deflection achieved.

It is shown in Appendix 6.B that the maximum deflection *for prescribed space dimension $N_{\mathcal{S}} = N$* is given by

$$\tilde{d}_{\max,N} = \frac{1}{\eta^2 N} \left(\sum_{k=1}^{N} \lambda_k^{(s)} \right)^2 . \qquad (6.11)$$

With[11]

$$\sum_{k=1}^{N} \lambda_k^{(s)} \leq N \lambda_1^{(s)} , \qquad \sum_{k=1}^{N} \lambda_k^{(s)} \leq \sum_{k=1}^{\infty} \lambda_k^{(s)} = \overline{E}_s ,$$

the following two upper bounds on $\tilde{d}_{\max,N}$ are obtained from (6.11):

$$\tilde{d}_{\max,N} \leq \frac{\lambda_1^{(s)2}}{\eta^2} N , \qquad \tilde{d}_{\max,N} \leq \frac{\overline{E}_s^2}{\eta^2} \frac{1}{N} . \qquad (6.12)$$

---

[11]Recall that $\lambda_1^{(s)}$ is the largest signal eigenvalue, and that $\lambda_k^{(s)} \geq 0$ since $\mathbf{R}_s$ is positive (semi-)definite.

For a hypothetical eigenvalue sequence $\lambda_k^{(s)} = e^{-ak^2}$ ($k = 1, 2, ...$), **Fig. 6.3** shows the dependence of $\tilde{d}_{\max,N}$ and of the two upper bounds (6.12) on N. The geometry of this dependence suggests that the point where the two bounds intersect may be taken as a crude estimate of the optimum dimension N_{opt} maximizing $\tilde{d}_{\max,N}$. This estimate is obtained as

$$\hat{N}_{\mathrm{opt}} = \mathrm{round}\left\{ \frac{\overline{E}_s}{\lambda_1^{(s)}} \right\}. \tag{6.13}$$

6.2.2 Two Special Cases

We briefly consider the two special situations where the signal $s(t)$ is stationary or nonstationary white (recall that the noise $n(t)$ has been assumed to be stationary and white anyway).

Stationary processes. For $s(t)$ and $n(t)$ jointly wide-sense stationary with power spectral densities $S_s(f)$ and $S_n(f) \equiv \eta$, respectively, it is reasonable to use a time-*invariant*[12] projection system $\mathbf{P}_S$. As in Section 6.1.2, we consider the stationary case as a limiting case of the nonstationary case:

$$k \longrightarrow f$$

$$\lambda_k^{(s)} \longrightarrow S_s(f)$$

$$u_k^{(s)}(t) \longrightarrow e^{j2\pi ft}$$

$$P_{S_{\mathrm{opt}}}(t_1, t_2) = \sum_{1 \leq k \leq N_{\mathrm{opt}}} u_k^{(s)}(t_1)\, u_k^{(s)*}(t_2) \longrightarrow \int_{f \in B} e^{j2\pi ft_1}\, e^{-j2\pi ft_2}\, df$$

$$= \int_f H(f)\, e^{j2\pi f(t_1 - t_2)}\, df.$$

Here, the projection system $\mathbf{P}_S$ is a time-invariant filter with binary-valued frequency response

---

[12] Our discussion of the stationary case is more concerned with interpretation (motivating the TF design to be considered in Section 6.2.3) than with implementation. The problem with a time-invariant projection system is that $Q_r = \|\mathbf{P}_S r\|^2 = \langle \mathbf{P}_S r, r \rangle$ diverges for stationary $r(t)$ due to the infinite time integration involved, and the numerator and denominator of $\tilde{d}$ are infinite. This problem can be resolved by an *a priori* use of finite time integrations (resulting in a time-varying projection system). In contrast, we simply consider the stationary case formally as a limiting case of the nonstationary case, using a time-invariant projection system. The problem of divergence could here be resolved, in an approximate manner, by an *a posteriori* truncation of the time integration in the detection statistic.

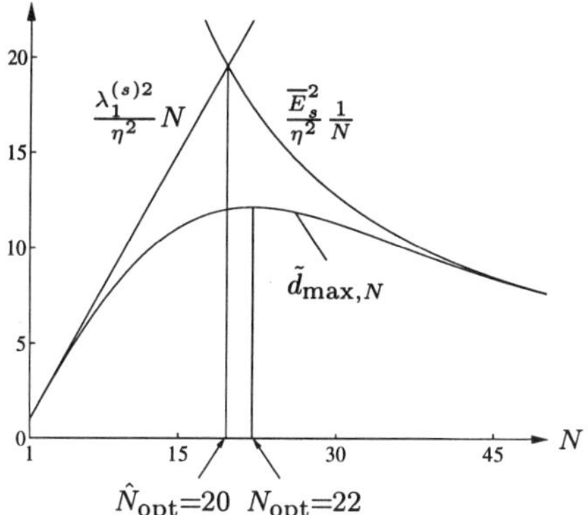

Figure 6.3. Dependence of $\tilde{d}_{\mathrm{max},N}$ and the two upper bounds $\frac{\lambda_1^{(s)2}}{\eta^2}N$ and $\frac{\overline{E}_s^2}{\eta^2}\frac{1}{N}$ on the space dimension N for a hypothetical eigenvalue sequence $\lambda_k^{(s)} = e^{-k^2/512}$ and $\eta = 1$.

$$H(f) = \begin{cases} 1 & \text{for } f \in B \\ 0 & \text{for } f \notin B, \end{cases}$$

where B is a (possibly multiple) pass band that will now be determined. In the nonstationary case, the index range $1 \le k \le N_{\mathrm{opt}}$ can be reformulated as $\lambda_k^{(s)} \ge \sigma$ with the eigenvalue threshold $\sigma = \lambda_{N_{\mathrm{opt}}}^{(s)}$ (recall that the $\lambda_k^{(s)}$ are arranged in non-increasing order). In the stationary case, this corresponds to defining the pass band B by $S_s(f) \ge \sigma$ with a suitably chosen threshold σ. As illustrated in **Fig. 6.4**, there is a one-to-one correspondence between the threshold σ (determining the band B) and the total bandwidth F of B. The optimum bandwidth F is the limit of the "eigenvalue bandwidth" (6.9),

$$N_{\mathrm{opt}} = \arg\max_N \left\{ \frac{1}{N} \left(\sum_{k=1}^N \lambda_k^{(s)} \right)^2 \right\},$$

which gives

$$F_{\mathrm{opt}} = \arg\max_F \left\{ \frac{1}{F} \left(\int_{f \in B} S_s(f)\, df \right)^2 \right\}.$$

This can be reformulated as an optimization of the threshold σ,

$$\sigma_{\mathrm{opt}} = \arg\max_\sigma \left\{ \frac{1}{F(\sigma)} \left(\int_{f \in B(\sigma)} S_s(f)\, df \right)^2 \right\},$$

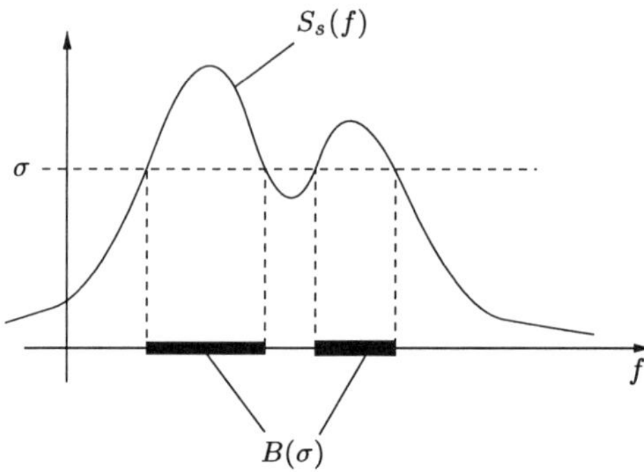

Figure 6.4.   Pass band of the projection system in the stationary case.

with $B(\sigma) = \{f : S_s(f) \geq \sigma\}$ and $F(\sigma)$ the bandwidth of $B(\sigma)$ (see Fig. 6.4).

Nonstationary white processes. In the dual case of a nonstationary white $s(t)$ with average instantaneous intensity $q_s(t)$ (see Section 6.1.2), an analogous discussion leads to a "frequency-invariant" projection system corresponding to a time gating,

$$\left(\mathbf{P}_S r\right)(t) = h(t)\, r(t) \qquad \text{with} \quad h(t) = \begin{cases} 1 & \text{for } t \in I(\sigma) \\ 0 & \text{for } t \notin I(\sigma) . \end{cases}$$

Here, $I(\sigma)$ is a (possibly multiple) time interval defined as $I(\sigma) = \{t : q_s(t) \geq \sigma\}$. An optimum value of the threshold σ is given by

$$\sigma_{\text{opt}} = \arg\max_\sigma \left\{ \frac{1}{T(\sigma)} \left(\int_{t \in I(\sigma)} q_s(t)\, dt \right)^2 \right\},$$

where $T(\sigma)$ is the total duration of $I(\sigma)$.

6.2.3   Time-Frequency Design

In the stationary case, the projection system was seen to pass all frequencies f where $S_s(f) \geq \sigma$, and in the nonstationary white case, it passes all times t where $q_s(t) \geq \sigma$. In analogy to the case of signal estimation (see Section 6.1.3), this suggests a heuristic *TF design* of the projection system for the general case of a nonstationary, nonwhite signal.

Let us define the TF region $R(\sigma)$ comprising all TF points where the WVS of the signal is larger than a threshold,

$$R(\sigma) \triangleq \{(t, f) : \ \overline{W}_s(t, f) \geq \sigma\} .$$

The threshold (and thus the TF region $R(\sigma)$) is chosen as

$$\sigma_0 \;=\; \arg\max_{\sigma} \left\{ \frac{1}{A(\sigma)} \left(\iint_{(t,f)\in R(\sigma)} \overline{W}_s(t,f)\,dt\,df \right)^{\!2} \right\},$$

where $A(\sigma)$ is the area of the TF region $R(\sigma)$. We now replace the optimum projection filter derived in Section 6.2.1 by the TF projection filter corresponding to $R(\sigma)$ in the sense of Chapter 5, i.e., the orthogonal projection operator $\mathbf{P}_{\mathcal{U}_{R(\sigma)}}$ on the eigenspace $\mathcal{U}_{R(\sigma)}$ of $R(\sigma)$. In analogy to the discussion given in Section 6.1.3, it can be argued that this heuristic TF projection filter will approximate the optimum projection filter if the signal process $s(t)$ is *underspread*. Indeed, for an underspread $s(t)$, the WVS $\overline{W}_s(t,f)$ will locally approximate the eigenvalues $\lambda_k^{(s)}$,

$$\overline{W}_s(t,f) \;=\; L_{\mathbf{R}_s}(t,f) \;\approx\; \lambda_k^{(s)} \qquad \text{for } (t,f)\in R_k\,,$$

where R_k is the TF region in which the eigenfunction $u_k^{(s)}(t)$ is located (see Section 6.1.3). Hence, including all eigenfunctions $u_k^{(s)}(t)$ with $\lambda_k^{(s)} \geq \sigma$ in the basis spanning the optimum space $\mathcal{S}_{\mathrm{opt}}$ is approximately equivalent to passing all the TF regions R_k in which $\overline{W}_s(t,f) \geq \sigma$.

6.2.4 Simulation Results

Fig. 6.5 and **Fig. 6.6** compare the optimum projection filter and TF projection filter for a signal process $s(t)$ whose WVS and eigenvalues are shown in Fig. 6.5(a) and (b), respectively. The process $s(t)$ was synthesized using the method proposed in [Hlawatsch and Kozek, 1995]. The dependence of $\tilde{d}_{\mathrm{max},N}$ and the upper bounds $\frac{\lambda_1^{(s)2}}{\eta^2}\,N$ and $\frac{\overline{E}_s^2}{\eta^2}\frac{1}{N}$ on N is depicted in Fig. 6.5(c). The dimension estimate $\hat{N}_{\mathrm{opt}}$ in (6.13) is seen to yield a reasonable result. Figs. 6.5(d) and (e) show that the WDs of the spaces $\mathcal{S}_{\mathrm{opt}}$ and $\mathcal{U}_{R(\sigma)}$ are very similar. Hence, the TF projection filter approximates the optimum projection filter quite well. Indeed, Fig. 6.6 shows that the optimum detector (using the optimum projection filter) and the TF designed detector (using the TF projection filter) have nearly identical performance. The deflections achieved were $\tilde{d}_{\mathrm{max}} = 9.716$ for the optimum detector and $\tilde{d} = 9.699$ for the TF designed detector.

6.2.5 Detection of Time-Frequency Localized Signals

We finally apply our results to the detection of a signal $s(t)$ that is known to be located within a given TF region R. The additive noise $n(t)$ is still assumed to

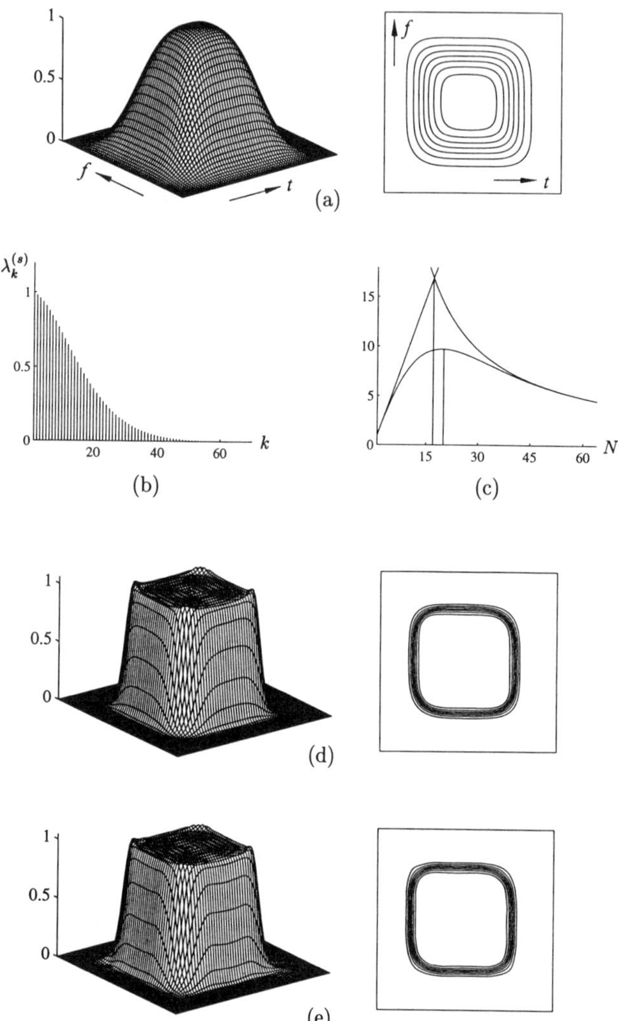

Figure 6.5. Comparison of optimum projection filter and TF projection filter for signal detection (I): (a) WVS of signal $s(t)$, (b) eigenvalues of $s(t)$, (c) dependence of $\tilde{d}_{\max,N}$ and the upper bounds $\frac{\lambda_1^{(s)2}}{\eta^2} N$ and $\frac{\overline{E}_s^2}{\eta^2} \frac{1}{N}$ on the space dimension N (cf. Fig. 6.3), (d) WD of space $\mathcal{S}_{\mathrm{opt}}$ underlying the optimum projection filter, and (e) WD of space $\mathcal{U}_{R(\sigma)}$ underlying the TF projection filter. (A slight smoothing has been applied to all WVS and WDs.)

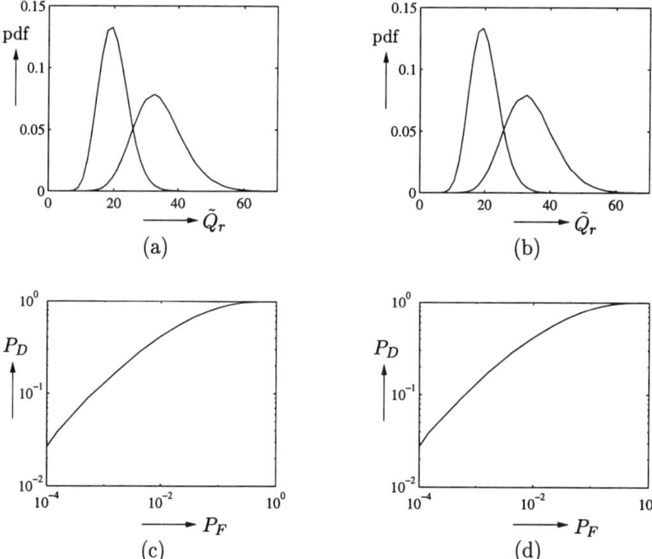

Figure 6.6. Comparison of optimum projection filter and TF projection filter for signal detection (II): (a) estimated conditional probability density functions of optimum detection statistic (i.e., quadratic form using optimum projection filter) under either hypothesis, (b) estimated conditional probability density functions of TF designed detection statistic (i.e., quadratic form using TF projection filter) under either hypothesis, (c) estimated receiver operator characteristics (ROC) [Van Trees, 1968] for optimum detection statistic, and (d) estimated ROC for TF designed detection statistic. These results were obtained by means of Monte Carlo simulation using 300000 realizations.

be stationary and white. As explained in Section 6.1.5, we model the signal as a nonstationary random process with autocorrelation operator $\mathbf{R}_s = \lambda^{(s)} \mathbf{P}_{\mathcal{U}_R}$, where $\lambda^{(s)}$ depends on the signal's energy (which need not be known however).

According to Section 6.2.1, the space underlying the optimum projection filter is spanned by the signal eigenfunctions $u_k^{(s)}(t)$ corresponding to the N_{opt} largest signal eigenvalues $\lambda_k^{(s)}$, with N_{opt} defined in (6.9). With $\mathbf{R}_s = \lambda^{(s)} \mathbf{P}_{\mathcal{U}_R}$, the signal eigenfunctions are $u_k^{(s)}(t) = u_k(t)$ where $\{u_k(t)\}_{k=1}^{N_R}$ is the eigensignal basis spanning the eigenspace $\mathcal{U}_R$, and the signal eigenvalues are identical for $k = 1, 2, ..., N_R$ and zero for $k > N_R$,

$$\lambda_k^{(s)} = \begin{cases} \lambda^{(s)} & \text{for } k = 1, 2, ..., N_R \\ 0 & \text{for } k > N_R. \end{cases}$$

It is then easily seen that for this eigenvalue distribution

$$\frac{1}{N}\left(\sum_{k=1}^{N}\lambda_k^{(s)}\right)^2 = \begin{cases} \lambda^{(s)2}\,N & \text{for } N = 1, 2, ..., N_R \\ \lambda^{(s)2}\,\frac{N_R^2}{N} & \text{for } N > N_R. \end{cases} \tag{6.14}$$

Since $\frac{N_R^2}{N} < N_R$ for $N > N_R$, the optimum space dimension is obtained as

$$N_{\text{opt}} = \arg\max_N\left\{\frac{1}{N}\left(\sum_{k=1}^{N}\lambda_k^{(s)}\right)^2\right\} = N_R. \tag{6.15}$$

Hence, the optimum space is simply the eigenspace of the signal region R,

$$\mathcal{S}_{\text{opt}} = \text{span}\{u_k(t)\}_{k=1}^{N_R} = \mathcal{U}_R,$$

and the optimum projection filter is simply the TF projection filter corresponding to R according to Chapter 5. This filter passes the entire TF region R in which the signal is known to be contained, and suppresses the rest of the TF plane in which there is only noise. Note that this simple solution does not depend on the average signal energy or the noise level. It coincides with the result of the approximate TF design discussed in Section 6.2.3 if the WVS of $s(t)$, $\overline{W}_s(t, f) = \lambda^{(s)}\,W_{\mathcal{U}_R}(t, f)$, is replaced by $\lambda^{(s)}I_R(t, f)$ as suggested in an analogous context in Section 6.1.5.

With (6.10), (6.14) and (6.15), the maximum deflection achieved is

$$\tilde{d}_{\max} = \tilde{d}\,\bigg|_{\mathcal{S}=\mathcal{S}_{\text{opt}}} = \frac{\lambda^{(s)2}N_R}{\eta^2} = \frac{E_s^2}{\eta^2 N_R},$$

where we have used that $\lambda^{(s)} = E_s/N_R$ (cf. Section 6.1.5).

Appendix 6.A: Proof of Theorem 6.1

Using the idempotency and self-adjointness of the projection operator $\mathbf{P}_\mathcal{S}$ (i.e., $\mathbf{P}_\mathcal{S}^2 = \mathbf{P}_\mathcal{S}$ and $\mathbf{P}_\mathcal{S}^+ = \mathbf{P}_\mathcal{S}$) as well as the uncorrelatedness of signal and noise (i.e., $R_{s,n}(t, t') = \text{E}\{s(t)\,n^*(t')\} \equiv 0$ or equivalently $\mathbf{R}_{s,n} = \mathbf{0}$), the mean error energy $\overline{E}_e$ can be developed as[13]

$$\overline{E}_e = \text{tr}\{\mathbf{R}_e\} = \text{tr}\{\mathbf{R}_{s-\mathbf{P}_\mathcal{S}r}\}$$

$$= \text{tr}\{\mathbf{R}_s\} - \text{tr}\{\mathbf{R}_{s,\mathbf{P}_\mathcal{S}r}\} - \text{tr}\{\mathbf{R}_{\mathbf{P}_\mathcal{S}r,s}\} + \text{tr}\{\mathbf{R}_{\mathbf{P}_\mathcal{S}r}\}$$

---

[13]Note that $\text{tr}\{\mathbf{AB}\} = \text{tr}\{\mathbf{BA}\}$.

$$
\begin{aligned}
&= \operatorname{tr}\{\mathbf{R}_s\} - \left[\operatorname{tr}\{\mathbf{R}_{s,s+n}\mathbf{P}_{\mathcal{S}}^+\} + \operatorname{tr}\{\mathbf{P}_{\mathcal{S}}\mathbf{R}_{s+n,s}\} - \operatorname{tr}\{\mathbf{P}_{\mathcal{S}}\mathbf{R}_{s+n}\mathbf{P}_{\mathcal{S}}^+\}\right] \\
&= \overline{E}_s - \left[\operatorname{tr}\{\mathbf{P}_{\mathcal{S}}^+(\mathbf{R}_s + \mathbf{R}_{s,n})\} + \operatorname{tr}\{\mathbf{P}_{\mathcal{S}}(\mathbf{R}_s + \mathbf{R}_{n,s})\}\right. \\
&\qquad \left. - \operatorname{tr}\{\mathbf{P}_{\mathcal{S}}\mathbf{P}_{\mathcal{S}}^+(\mathbf{R}_s + \mathbf{R}_{s,n} + \mathbf{R}_{n,s} + \mathbf{R}_n)\}\right] \\
&= \overline{E}_s - \operatorname{tr}\{\mathbf{P}_{\mathcal{S}}(\mathbf{R}_s + \mathbf{R}_s - \mathbf{R}_s - \mathbf{R}_n)\},
\end{aligned}
$$

which finally gives

$$
\overline{E}_e = \overline{E}_s - \operatorname{tr}\{\mathbf{P}_{\mathcal{S}}\mathbf{D}\} \qquad \text{with} \quad \mathbf{D} = \mathbf{R}_s - \mathbf{R}_n. \tag{6.A.1}
$$

Hence, minimization of $\overline{E}_e$ is equivalent to maximization of $\operatorname{tr}\{\mathbf{P}_{\mathcal{S}}\mathbf{D}\}$,

$$
\mathcal{S}_{\mathrm{opt}} = \arg\max_{\mathcal{S}} \operatorname{tr}\{\mathbf{P}_{\mathcal{S}}\mathbf{D}\}.
$$

Let $\{s_k(t)\}_{k=1}^{N_{\mathcal{S}}}$ be an orthonormal basis of the looked-for space $\mathcal{S}$. The kernel of $\mathbf{P}_{\mathcal{S}}$ can then be written as

$$
P_{\mathcal{S}}(t,t') = \sum_{k=1}^{N_{\mathcal{S}}} s_k(t)\, s_k^*(t'),
$$

and $\operatorname{tr}\{\mathbf{P}_{\mathcal{S}}\mathbf{D}\}$ becomes

$$
\begin{aligned}
\operatorname{tr}\{\mathbf{P}_{\mathcal{S}}\mathbf{D}\} &= \int_t \int_{t'} P_{\mathcal{S}}(t,t')\, D(t',t)\, dt\, dt' \\
&= \sum_{k=1}^{N_{\mathcal{S}}} \left[\int_t \int_{t'} D(t',t)\, s_k(t)\, s_k^*(t')\, dt\, dt'\right] \\
&= \sum_{k=1}^{N_{\mathcal{S}}} \langle \mathbf{D}s_k,\, s_k \rangle.
\end{aligned}
$$

With this formulation, the results of Theorem 2.1 (see Section 2.7.1) are directly applicable. Hence, the optimum basis signals $s_k(t)$ are the eigenfunctions of the operator $\mathbf{D}$ corresponding to all positive eigenvalues. Denoting these eigenvalues and eigenfunctions as $\lambda_k^{(D)}$ and $u_k^{(D)}(t)$ for $k = 1, ..., N_+$, with N_+ the number of positive eigenvalues $\lambda_k^{(D)}$, it follows from Theorem 2.1 that

$$
\left.\operatorname{tr}\{\mathbf{P}_{\mathcal{S}}\mathbf{D}\}\right|_{\mathrm{max}} = \sum_{k=1}^{N_+} \lambda_k^{(D)}.
$$

Inserting this into (6.A.1) yields (6.1).

Appendix 6.B: Proof of Theorem 6.2

We start by expressing the individual terms of the deflection $\tilde{d}$ in (6.8). The terms in the numerator of $\tilde{d}$ are

$$\mathrm{E}_0\{\tilde{Q}_r\} = \mathrm{E}\{\|\mathbf{P}_{\mathcal{S}}\,n\|^2\} = \mathrm{E}\{\langle \mathbf{P}_{\mathcal{S}}\,n, n\rangle\}$$

$$= \mathrm{E}\left\{\int_t \int_{t'} P_{\mathcal{S}}(t,t')\,n(t')\,n^*(t)\,dt\,dt'\right\}$$

$$= \int_t \int_{t'} P_{\mathcal{S}}(t,t')\,R_n(t',t)\,dt\,dt'$$

$$= \mathrm{tr}\{\mathbf{P}_{\mathcal{S}}\,\mathbf{R}_n\}\,, \tag{6.B.1}$$

and, in a similar manner,

$$\mathrm{E}_1\{\tilde{Q}_r\} = \mathrm{tr}\{\mathbf{P}_{\mathcal{S}}\,\mathbf{R}_{s+n}\} = \mathrm{tr}\{\mathbf{P}_{\mathcal{S}}\,\mathbf{R}_s\} + \mathrm{tr}\{\mathbf{P}_{\mathcal{S}}\,\mathbf{R}_n\}\,. \tag{6.B.2}$$

The denominator of $\tilde{d}$ is

$$\mathrm{var}_0\{\tilde{Q}_r\} = \mathrm{E}_0\{\tilde{Q}_r^2\} - \left(\mathrm{E}_0\{\tilde{Q}_r\}\right)^2 \tag{6.B.3}$$

with

$$\mathrm{E}_0\{\tilde{Q}_r^2\} = \mathrm{E}\{\langle \mathbf{P}_{\mathcal{S}}\,n, n\rangle \langle \mathbf{P}_{\mathcal{S}}\,n, n\rangle^*\}$$

$$= \int_{t_1} \int_{t_1'} \int_{t_2} \int_{t_2'} P_{\mathcal{S}}(t_1,t_1')\,P_{\mathcal{S}}^*(t_2,t_2')\,\mathrm{E}\{n(t_1')\,n^*(t_1)\,n^*(t_2')\,n(t_2)\}\,dt_1\,dt_1'\,dt_2\,dt_2'$$

$$= \int_{t_1} \int_{t_1'} \int_{t_2} \int_{t_2'} P_{\mathcal{S}}(t_1,t_1')\,P_{\mathcal{S}}^*(t_2,t_2')$$

$$\cdot \left[R_n(t_1',t_1)\,R_n(t_2,t_2') + R_n(t_1',t_2')\,R_n(t_2,t_1)\right]\,dt_1\,dt_1'\,dt_2\,dt_2'$$

$$= \left[\int_{t_1} \int_{t_1'} P_{\mathcal{S}}(t_1,t_1')\,R_n(t_1',t_1)\,dt_1\,dt_1'\right]\left[\int_{t_2} \int_{t_2'} P_{\mathcal{S}}^*(t_2,t_2')\,R_n(t_2,t_2')\,dt_2\,dt_2'\right]$$

$$+ \int_{t_1} \int_{t_1'} \int_{t_2} \int_{t_2'} P_{\mathcal{S}}(t_1,t_1')\,R_n(t_1',t_2')\,P_{\mathcal{S}}^*(t_2,t_2')\,R_n(t_2,t_1)\,dt_1\,dt_1'\,dt_2\,dt_2'$$

$$= \mathrm{tr}\{\mathbf{P}_{\mathcal{S}}\,\mathbf{R}_n\}\,\mathrm{tr}\{\mathbf{P}_{\mathcal{S}}^+\,\mathbf{R}_n\} + \mathrm{tr}\{\mathbf{P}_{\mathcal{S}}\,\mathbf{R}_n\,\mathbf{P}_{\mathcal{S}}^+\,\mathbf{R}_n\}$$

$$= \left(\mathrm{tr}\{\mathbf{P}_{\mathcal{S}}\,\mathbf{R}_n\}\right)^2 + \mathrm{tr}\{(\mathbf{P}_{\mathcal{S}}\,\mathbf{R}_n)^2\}\,, \tag{6.B.4}$$

where we have used that

$$\mathrm{E}\{n(t_1')\,n^*(t_1)\,n^*(t_2')\,n(t_2)\} = R_n(t_1',t_1)\,R_n(t_2,t_2') + R_n(t_1',t_2')\,R_n(t_2,t_1)$$

due to the Gaussianity and circular symmetry of $n(t)$ [Therrien, 1992]. Inserting (6.B.1)–(6.B.4) into (6.8), the deflection is obtained as

$$\tilde{d} = \frac{(\mathrm{tr}\{\mathbf{P}_{\mathcal{S}}\mathbf{R}_s\})^2}{\mathrm{tr}\{(\mathbf{P}_{\mathcal{S}}\mathbf{R}_n)^2\}} \cdot$$

Maximization of this expression with respect to the space $\mathcal{S}$ appears to be difficult in general. Therefore, we now restrict our development to the special case where the noise $n(t)$ is stationary and white with power spectral density η. With $\mathbf{R}_n = \eta\mathbf{I}$, the deflection here simplifies to

$$\tilde{d} = \frac{(\mathrm{tr}\{\mathbf{P}_{\mathcal{S}}\mathbf{R}_s\})^2}{\mathrm{tr}\{(\mathbf{P}_{\mathcal{S}}\eta\mathbf{I})^2\}} = \frac{(\mathrm{tr}\{\mathbf{P}_{\mathcal{S}}\mathbf{R}_s\})^2}{\eta^2\,\mathrm{tr}\{\mathbf{P}_{\mathcal{S}}\}} = \frac{(\mathrm{tr}\{\mathbf{P}_{\mathcal{S}}\mathbf{R}_s\})^2}{\eta^2 N_{\mathcal{S}}}, \tag{6.B.5}$$

where $N_{\mathcal{S}} = \mathrm{tr}\{\mathbf{P}_{\mathcal{S}}\}$ is the dimension of $\mathcal{S}$.

Let us first assume a fixed dimension $N_{\mathcal{S}} = N$. The optimum space is then

$$\mathcal{S}_{\mathrm{opt}}(N) = \arg\max_{N_{\mathcal{S}}=N} \tilde{d} = \arg\max_{N_{\mathcal{S}}=N} \left(\mathrm{tr}\{\mathbf{P}_{\mathcal{S}}\mathbf{R}_s\}\right)^2.$$

It is easily shown that (cf. Appendix 6.A)

$$\mathrm{tr}\{\mathbf{P}_{\mathcal{S}}\mathbf{R}_s\} = \sum_{k=1}^{N_{\mathcal{S}}} \langle \mathbf{R}_s s_k, s_k \rangle,$$

where $\{s_k(t)\}_{k=1}^{N_{\mathcal{S}}}$ is an orthonormal basis of $\mathcal{S}$. Since $\mathbf{R}_s$ is positive (semi-) definite, $\mathrm{tr}\{\mathbf{P}_{\mathcal{S}}\mathbf{R}_s\} \geq 0$. Hence, maximizing $(\mathrm{tr}\{\mathbf{P}_{\mathcal{S}}\mathbf{R}_s\})^2$ is equivalent to maximizing $\mathrm{tr}\{\mathbf{P}_{\mathcal{S}}\mathbf{R}_s\}$, and $\mathcal{S}_{\mathrm{opt}}(N)$ can equivalently be formulated as

$$\mathcal{S}_{\mathrm{opt}}(N) = \arg\max_{N_{\mathcal{S}}=N} \mathrm{tr}\{\mathbf{P}_{\mathcal{S}}\mathbf{R}_s\} = \arg\max \sum_{k=1}^{N} \langle \mathbf{R}_s s_k, s_k \rangle.$$

At this point, the results of Theorem 2.1 (see Section 2.7.1) are directly applicable. Hence, the optimum basis signals $s_k(t)$ are the eigenfunctions $u_k^{(s)}(t)$ of the signal autocorrelation operator $\mathbf{R}_s$ corresponding to the N largest eigenvalues $\lambda_k^{(s)}$ (note that $\lambda_k^{(s)} \geq 0$). Assuming the $\lambda_k^{(s)}$ to be arranged in non-increasing order, the N largest eigenvalues are the first N eigenvalues and we obtain

$$\mathcal{S}_{\mathrm{opt}}(N) = \mathrm{span}\{u_k^{(s)}(t)\}_{k=1}^{N}.$$

Furthermore, it follows from Theorem 2.1 that

$$\mathrm{tr}\{\mathbf{P}_{\mathcal{S}}\mathbf{R}_s\}\Big|_{\mathrm{max},N} = \sum_{k=1}^{N} \lambda_k^{(s)}.$$

Inserting into (6.B.5), the maximum deflection (for given space dimension $N_\mathcal{S} = N$) is obtained as

$$\tilde{d}_{\mathrm{max},N} \;=\; \frac{1}{\eta^2 N}\left(\sum_{k=1}^{N}\lambda_k^{(s)}\right)^{2}. \qquad (6.\mathrm{B}.6)$$

Finally, with (6.B.6) the optimum space dimension N becomes

$$N_{\mathrm{opt}} \;=\; \arg\max_{N}\left\{\frac{1}{N}\left(\sum_{k=1}^{N}\lambda_k^{(s)}\right)^{2}\right\},$$

and the maximum deflection is hence obtained as

$$\tilde{d}_{\mathrm{max}} \;=\; \frac{1}{\eta^2 N_{\mathrm{opt}}}\left(\sum_{k=1}^{N_{\mathrm{opt}}}\lambda_k^{(s)}\right)^{2}.$$

7   THE AMBIGUITY FUNCTION OF A LINEAR SIGNAL SPACE

For TF signal analysis, two fundamental types of TF signal representations can be distingushed, namely, the "energetic" type and the "correlative" type [Hlawatsch, 1991]. The energetic mode of TF analysis seeks to define "TF energy distributions" that satisfy the marginal properties (1.6)–(1.8) or, more generally, give an indication of how the signal is distributed over the TF plane. Two prominent energetic TF distributions are the WD and the spectrogram.

The correlative mode of TF analysis, on the other hand, seeks to define joint TF correlation functions. The central correlative TF representation is the *ambiguity function* (AF) that has been briefly reviewed in Section 1.2. The cross-AF of two signals $x(t)$ and $y(t)$ is defined as

$$A_{x,y}(\tau, \nu) \triangleq \int_t x\left(t + \frac{\tau}{2}\right) y^*\left(t - \frac{\tau}{2}\right) e^{-j2\pi\nu t} \, dt, \qquad (7.1)$$

and the auto-AF of a single signal $x(t)$ is defined as $A_x(\tau, \nu) \triangleq A_{x,x}(\tau, \nu)$. The AF is a joint TF correlation function with many interesting properties and applications [Woodward, 1953, Rihaczek, 1969, Van Trees, 1992, Hlawatsch and Boudreaux-Bartels, 1992, Hlawatsch and Flandrin, 1997, Szu and Blodgett, 1981]. In particular, the application of the AF to the radar/sonar problem of jointly estimating the range and radial velocity (Doppler shift) of a slowly fluctuating point target will be reviewed in Section 8.1.

This work has so far been concerned exclusively with the energetic TF analysis of linear signal spaces. In this chapter and the next one, however, we shall consider the correlative TF analysis of linear signal spaces.

The organization of this chapter is largely parallel to that of Chapter 2. In Section 7.1, we introduce the *AF of a linear signal space*. Section 7.2 investigates basic properties of the AF of a linear signal space. The results obtained for some important special signal spaces are discussed in Section 7.3. Section 7.4 presents two interpretations of the AF, one related to sophisticated spaces (cf. Section 3.1.2) and the other related to the statistical TF correlation of random processes. Finally, the cross-AF of two spaces and a discrete-time AF version are introduced in Section 7.5.

The application of the AF of a linear signal space to the analysis of a "multipulse" range-Doppler estimator will be discussed in Chapter 8.

7.1 Definitions and Expressions

Based on the AF of a signal in (7.1), the *AF of a linear signal space* $\mathcal{X}$ [Hlawatsch and Edelson, 1992] can be defined using the general principle of defining quadratic space representations (see Section 2.2); this yields

$$A_{\mathcal{X}}(\tau,\nu) \;=\; \int_t P_{\mathcal{X}}\left(t+\frac{\tau}{2}, t-\frac{\tau}{2}\right) e^{-j2\pi\nu t}\, dt \tag{7.2}$$

$$=\; \sum_{k=1}^{N_{\mathcal{X}}} A_{x_k}(\tau,\nu)\,, \tag{7.3}$$

where $P_{\mathcal{X}}(t,t')$ is the kernel of the projection operator $\mathbf{P}_{\mathcal{X}}$ and $\{x_k(t)\}_{k=1}^{N_{\mathcal{X}}}$ is an arbitrary orthonormal basis of $\mathcal{X}$. We see that the AF of a linear signal space is simply the sum of the AFs of all orthonormal basis signals, which is independent of the specific orthonormal basis of $\mathcal{X}$. Introducing the basis vector $\mathbf{x}(t) = [x_1(t), x_2(t), ..., x_{N_{\mathcal{X}}}(t)]^T$, the last expression can be written as

$$A_{\mathcal{X}}(\tau,\nu) \;=\; \int_t \mathbf{x}^H\left(t-\frac{\tau}{2}\right) \mathbf{x}\left(t+\frac{\tau}{2}\right) e^{-j2\pi\nu t}\, dt \,,$$

which is reminiscent of the AF of a signal.

The projection operator $\mathbf{P}_{\mathcal{X}}$ can be recovered from $A_{\mathcal{X}}(\tau,\nu)$ according to

$$P_{\mathcal{X}}(t_1,t_2) \;=\; \int_\nu A_{\mathcal{X}}\left(t_1-t_2,\nu\right) e^{j\pi(t_1+t_2)\nu}\, d\nu \,, \tag{7.4}$$

which shows that the AF of a signal space provides a *complete* characterization of the space.

A "frequency-domain" expression of $A_{\mathcal{X}}(\tau, \nu)$ in terms of the bifrequency function $\tilde{P}_{\mathcal{X}}(f, f')$ of $\mathbf{P}_{\mathcal{X}}$, as defined in (2.6), is

$$A_{\mathcal{X}}(\tau, \nu) = \int_f \tilde{P}_{\mathcal{X}}\left(f + \frac{\nu}{2}, f - \frac{\nu}{2}\right) e^{j2\pi\tau f} \, df.$$

Even though the AF of a signal space is the sum of the AFs of all (orthonormal) basis signals, it usually does not occupy a larger region in the (τ, ν)-plane than the AF of an individual signal. This is an important difference from the WD. Indeed, the AFs of two orthogonal signals are usually not located in different regions of the (τ, ν)-plane. Rather, they are both located about the origin of the (τ, ν)-plane, often with oscillatory "sidelobes" away from the origin. For *simple* spaces (cf. Section 3.1.2), these sidelobes tend to cancel when the AFs of the basis signals are added; the space's AF will then be even more concentrated about the origin of the (τ, ν)-plane than the AF of any individual basis signal.

In this context, it is important to note that the AF of a space is the 2-D Fourier transform of the WD of a space (cf. (2.28)),

$$A_{\mathcal{X}}(\tau, \nu) = \int_t \int_f W_{\mathcal{X}}(t, f) \, e^{-j2\pi(\nu t - \tau f)} \, dt \, df. \tag{7.5}$$

Hence, the broader and smoother the WD is in the (t, f)-plane, the more concentrated will the AF be about the origin of the (τ, ν)-plane. Evidently, the AF of a *simple, high-dimensional* space will be particularly concentrated: the space's simplicity guarantees that the WD is reasonably smooth, and the high dimension implies a large WD support. This issue will be pursued in Sections 7.4.1 and 8.2.

7.2 Properties

Since the AF of a space is the 2-D Fourier transform of the space's WD, most of the properties of the AF are dual (in the sense of the Fourier transform) to properties of the WD. In accordance with this duality, the following discussion of properties of the AF is largely parallel to Section 2.3.

1) Symmetry. The AF of a linear signal space is a complex-valued function that is symmetric about the origin,

$$A_{\mathcal{X}}^*(-\tau, -\nu) = A_{\mathcal{X}}(\tau, \nu).$$

2) Maximum. The maximum of the AF's magnitude occurs at the origin of the (τ, ν)-plane and equals the dimension (energy) of the space,

$$|A_{\mathcal{X}}(\tau, \nu)| \leq A_{\mathcal{X}}(0, 0) = N_{\mathcal{X}} = E_{\mathcal{X}}.$$

3) Norm. The squared norm of $A_{\mathcal{X}}(\tau, \nu)$ equals the dimension of $\mathcal{X}$,

$$\|A_{\mathcal{X}}\|^2 = \int_\tau \int_\nu |A_{\mathcal{X}}(\tau, \nu)|^2 \, d\tau \, d\nu = N_{\mathcal{X}} \, .$$

This shows that $A_{\mathcal{X}}(\tau, \nu)$ is square-integrable if and only if $\mathcal{X}$ has finite dimension. Combining properties 2) and 3) gives the identity

$$\int_\tau \int_\nu |A_{\mathcal{X}}(\tau, \nu)|^2 \, d\tau \, d\nu = \frac{|A_{\mathcal{X}}(0,0)|^2}{N_{\mathcal{X}}} \, , \tag{7.6}$$

which shows that the volume under $|A_{\mathcal{X}}(\tau, \nu)|^2$ equals the peak value $|A_{\mathcal{X}}(0,0)|^2$ divided by $N_{\mathcal{X}}$. This generalizes the "radar uncertainty principle" related to the AF of a signal (see Sections 8.1.3 and 8.2.4).

4) Finite support. If all signals $x(t) \in \mathcal{X}$ are time-limited to $[t_1, t_2]$ (i.e., $\mathcal{X}$ is a subspace of the space $\mathcal{T}[t_1, t_2]$ of all signals time-limited to $[t_1, t_2]$), then the AF of $\mathcal{X}$ is limited with respect to the time-lag parameter τ,

$$\mathcal{X} \subseteq \mathcal{T}[t_1, t_2] \quad \Rightarrow \quad A_{\mathcal{X}}(\tau, \nu) = 0 \quad \text{for } |\tau| > t_2 - t_1 \, .$$

Similarly, if all signals $x(t) \in \mathcal{X}$ are band-limited to a frequency band $[f_1, f_2]$ (i.e., $\mathcal{X}$ is a subspace of the space $\mathcal{F}[f_1, f_2]$ of all signals band-limited to $[f_1, f_2]$), then the AF is limited with respect to the frequency-lag parameter ν,

$$\mathcal{X} \subseteq \mathcal{F}[f_1, f_2] \quad \Rightarrow \quad A_{\mathcal{X}}(\tau, \nu) = 0 \quad \text{for } |\nu| > f_2 - f_1 \, .$$

5) Moyal-type relation I. The inner product of the AFs of two signal spaces $\mathcal{X}$ and $\mathcal{Y}$,

$$\langle A_{\mathcal{X}}, A_{\mathcal{Y}} \rangle = \int_\tau \int_\nu A_{\mathcal{X}}(\tau, \nu) \, A_{\mathcal{Y}}^*(\tau, \nu) \, d\tau \, d\nu \, ,$$

can be expressed in terms of the spaces' orthogonal projection operators $\mathbf{P}_{\mathcal{X}}$ and $\mathbf{P}_{\mathcal{Y}}$ or the spaces' orthonormal bases $\{x_k(t)\}_{k=1}^{N_{\mathcal{X}}}$ and $\{y_l(t)\}_{l=1}^{N_{\mathcal{Y}}}$ as follows:

$$\langle A_{\mathcal{X}}, A_{\mathcal{Y}} \rangle = \int_t \int_{t'} P_{\mathcal{X}}(t, t') \, P_{\mathcal{Y}}^*(t, t') \, dt \, dt' = \sum_{k=1}^{N_{\mathcal{X}}} \sum_{l=1}^{N_{\mathcal{Y}}} |\langle x_k, y_l \rangle|^2 \, .$$

This can be considered a generalization of Moyal's formula (1.11). Note also that

$$\langle A_{\mathcal{X}}, A_{\mathcal{Y}} \rangle = \langle W_{\mathcal{X}}, W_{\mathcal{Y}} \rangle \, .$$

It can be shown that $\langle A_{\mathcal{X}}, A_{\mathcal{Y}} \rangle$ is bounded as

$$0 \le \langle A_{\mathcal{X}}, A_{\mathcal{Y}} \rangle \le \min\{N_{\mathcal{X}}, N_{\mathcal{Y}}\} \left(\le \sqrt{N_{\mathcal{X}} N_{\mathcal{Y}}} \right) \, .$$

The lower bound is attained if and only if $\mathcal{X}$ and $\mathcal{Y}$ are orthogonal,

$$\langle A_\mathcal{X}, A_\mathcal{Y} \rangle = 0 \qquad \Longleftrightarrow \qquad \mathcal{X} \perp \mathcal{Y} .$$

The upper bound is attained if and only if one space is a subspace of the other space. For example, if $N_\mathcal{Y} \leq N_\mathcal{X}$ so that $\min\{N_\mathcal{X}, N_\mathcal{Y}\} = N_\mathcal{Y}$, we have

$$\langle A_\mathcal{X}, A_\mathcal{Y} \rangle = N_\mathcal{Y} \qquad \Longleftrightarrow \qquad \mathcal{Y} \subseteq \mathcal{X} .$$

6) Moyal-type relation II. The inner product of the AF of a signal $s(t)$ and the AF of a signal space $\mathcal{X}$ equals the energy of the projected signal,

$$\langle A_s, A_\mathcal{X} \rangle \;=\; \|s_\mathcal{X}\|^2 \;=\; \sum_{k=1}^{N_\mathcal{X}} |\langle s, x_k \rangle|^2 .$$

This can again be considered a generalization of Moyal's formula (1.11). It follows that $\langle A_s, A_\mathcal{X} \rangle$ is bounded as

$$0 \leq \langle A_s, A_\mathcal{X} \rangle \leq \|s\|^2 ,$$

where the lower bound is attained if and only if $s(t)$ is orthogonal to $\mathcal{X}$,

$$\langle A_s, A_\mathcal{X} \rangle = 0 \qquad \Longleftrightarrow \qquad s \perp \mathcal{X} ,$$

and the upper bound is attained if and only if $s(t)$ is an element of $\mathcal{X}$,

$$\langle A_s, A_\mathcal{X} \rangle = \|s\|^2 \qquad \Longleftrightarrow \qquad s \in \mathcal{X} .$$

7) Direct sum. If two spaces $\mathcal{X}$ and $\mathcal{Y}$ are orthogonal, then the AF of their (direct) sum equals the sum of the AFs of $\mathcal{X}$ and $\mathcal{Y}$,

$$\mathcal{X} \perp \mathcal{Y} \qquad \Rightarrow \qquad A_{\mathcal{X}+\mathcal{Y}}(\tau,\nu) = A_\mathcal{X}(\tau,\nu) + A_\mathcal{Y}(\tau,\nu) .$$

In general, the AF of the sum of two non-orthogonal spaces $\mathcal{X}$ and $\mathcal{Y}$ cannot be expressed in terms of $A_\mathcal{X}(\tau,\nu)$ and $A_\mathcal{Y}(\tau,\nu)$ in a simple way.

8) Derivation of other quadratic space representations. Any quadratic space representation (cf. Section 2.2)

$$Q_\mathcal{X}(\Theta) = \int_{t_1} \int_{t_2} k_Q(\Theta; t_1, t_2)\, P_\mathcal{X}(t_1, t_2)\, dt_1 dt_2$$

can be derived from the AF $A_\mathcal{X}(\tau,\nu)$ via a linear transformation,

$$Q_\mathcal{X}(\Theta) = \int_\tau \int_\nu \tilde{L}_Q(\Theta; \tau,\nu)\, A_\mathcal{X}(\tau,\nu)\, d\tau\, d\nu$$

where

$$\tilde{L}_Q(\Theta;\tau,\nu) = \int_t k_Q\left(\Theta;\, t+\frac{\tau}{2},\, t-\frac{\tau}{2}\right) e^{j2\pi\nu t}\, dt\,.$$

This transformation relating $A_{\mathcal{X}}(\tau,\nu)$ and $Q_{\mathcal{X}}(\Theta)$ is identical to the transformation relating the corresponding *signal* representations $A_x(\tau,\nu)$ and $Q_x(\Theta)$ [Hlawatsch, 1992a]. Therefore, any linear relation connecting the AF of a signal with some other quadratic signal representation can immediately be reformulated for a space. Application of this principle to the signal/space representations considered in Section 2.2.2 yields the following relations:

$$E_{\mathcal{X}} = A_{\mathcal{X}}(0,0)$$

$$d_{\mathcal{X}}(t) = \int_\nu A_{\mathcal{X}}(0,\nu)\, e^{j2\pi t\nu}\, d\nu$$

$$D_{\mathcal{X}}(f) = \int_\tau A_{\mathcal{X}}(\tau,0)\, e^{-j2\pi f\tau}\, d\tau$$

$$r_{\mathcal{X}}(\tau) = A_{\mathcal{X}}(\tau,0)$$

$$R_{\mathcal{X}}(\nu) = A_{\mathcal{X}}(0,\nu)$$

$$m_{\mathcal{X}}^{(n)} = \left(-\frac{1}{2\pi j}\right)^n \frac{\partial^n}{\partial\nu^n} A_{\mathcal{X}}(0,\nu)\Big|_{\nu=0}$$

$$M_{\mathcal{X}}^{(n)} = \left(\frac{1}{2\pi j}\right)^n \frac{\partial^n}{\partial\tau^n} A_{\mathcal{X}}(\tau,0)\Big|_{\tau=0}$$

$$W_{\mathcal{X}}(t,f) = \int_\tau\int_\nu A_{\mathcal{X}}(\tau,\nu)\, e^{j2\pi(t\nu-f\tau)}\, d\tau\, d\nu$$

$$S_{\mathcal{X}}^{(h)}(t,f) = \int_\tau\int_\nu A_h^*(\tau,\nu)\, A_{\mathcal{X}}(\tau,\nu)\, e^{j2\pi(t\nu-f\tau)}\, d\tau\, d\nu$$

$$C_{\mathcal{X}}^{(h)}(t,f) = \int_\tau\int_\nu A_h^*\left(\frac{f}{f_r}\tau,\, \frac{f_r}{f}\nu\right) A_{\mathcal{X}}(\tau,\nu)\, e^{j2\pi t\nu}\, d\tau\, d\nu\,.$$

In particular, we see that the temporal correlation $r_{\mathcal{X}}(\tau)$ and the spectral correlation $R_{\mathcal{X}}(\nu)$ are the AF slices along the τ-axis ($\nu=0$) and the ν-axis ($\tau=0$), respectively; these relations are the "marginal properties" of the AF.

9) Linear space transformations—General. Let $\mathbf{H}$ denote a linear signal/space transformation as considered in Section 2.3. If the operator $\mathbf{H}$ is unitary on $\mathcal{L}_2(\mathbb{R})$, then the AF of the transformed space $\mathbf{H}\mathcal{X}$ can be derived from the AF of $\mathcal{X}$ via a linear transformation,

$$A_{\mathbf{H}\mathcal{X}}(\tau,\nu) = \int_{\tau'}\int_{\nu'} \tilde{L}_H(\tau,\nu;\tau',\nu')\, A_{\mathcal{X}}(\tau',\nu')\, d\tau'd\nu'$$

with

$$\tilde{L}_H(\tau,\nu;\tau',\nu') \;=\; \int_t\!\!\int_{t'} H\!\left(t+\frac{\tau}{2},t'+\frac{\tau'}{2}\right) H^*\!\left(t-\frac{\tau}{2},t'-\frac{\tau'}{2}\right) e^{-j2\pi(\nu t-\nu' t')}\,dt\,dt'.$$

This linear transformation relating the AFs of the spaces $\mathcal{X}$ and $\mathbf{H}\mathcal{X}$ is identical to the linear transformation relating the AFs of the signals $x(t)$ and $(\mathbf{H}x)(t)$ [Hlawatsch, 1992a]. Thus, if the unitarity condition is met, then a relation known to hold in the signal case can immediately be reformulated for spaces. Two important examples of this principle are studied in the following.

10) Time-frequency coordinate transforms. Let $\mathbf{H}$ be the unitary operator corresponding to the affine, area-preserving TF coordinate transform

$$(t,f) \longrightarrow (\alpha t + \beta f - \tau_0,\; \gamma t + \delta f - \nu_0) \qquad \text{with } D = \alpha\delta - \gamma\beta = 1\,,$$

in the sense that the WD of a space is transformed as (cf. Section 2.3)

$$W_{\mathbf{H}\mathcal{X}}(t,f) \;=\; W_{\mathcal{X}}(\alpha t + \beta f - \tau_0,\; \gamma t + \delta f - \nu_0)\,.$$

With (7.5), it is easily shown that this implies the AF transformation

$$A_{\mathbf{H}\mathcal{X}}(\tau,\nu) \;=\; A_{\mathcal{X}}(\alpha\tau + \beta\nu,\; \gamma\tau + \delta\nu)\, e^{-j2\pi[\tau_0(\gamma\tau+\delta\nu)-\nu_0(\alpha\tau+\beta\nu)]}\,,$$

which, up to a phase factor, amounts to the linear coordinate transform $(\tau,\nu) \rightarrow (\alpha\tau + \beta\nu,\; \gamma\tau + \delta\nu)$. Apart from the missing shift component (τ_0 and ν_0), this is analogous to the original coordinate transform in the (t,f)-plane. Some special cases of particular importance are listed below.

TF shift:

$$(\mathbf{H}x)(t) = x(t-\tau_0)\, e^{j2\pi\nu_0 t} \qquad\Rightarrow\qquad A_{\mathbf{H}\mathcal{X}}(\tau,\nu) = A_{\mathcal{X}}(\tau,\nu)\, e^{-j2\pi(\tau_0\nu-\nu_0\tau)}\,.$$

TF scaling:

$$(\mathbf{H}x)(t) = \sqrt{|a|}\, x(at) \qquad\Rightarrow\qquad A_{\mathbf{H}\mathcal{X}}(\tau,\nu) = A_{\mathcal{X}}\!\left(a\tau,\frac{\nu}{a}\right).$$

Chirp convolution:

$$(\mathbf{H}x)(t) = x(t) * \sqrt{|c|}\, e^{j\pi c t^2} \qquad\Rightarrow\qquad A_{\mathbf{H}\mathcal{X}}(\tau,\nu) = A_{\mathcal{X}}\!\left(\tau-\frac{\nu}{c},\nu\right). \qquad (7.7)$$

Chirp multiplication:

$$(\mathbf{H}x)(t) = x(t)\, e^{j\pi c t^2} \qquad\Rightarrow\qquad A_{\mathbf{H}\mathcal{X}}(\tau,\nu) = A_{\mathcal{X}}(\tau,\nu-c\tau)\,. \qquad (7.8)$$

Fourier transform:

$$(\mathbf{H}x)(t) = \sqrt{|c|}\, X(ct) \quad \Rightarrow \quad A_{\mathbf{H}\mathcal{X}}(\tau,\nu) = A_{\mathcal{X}}\left(-\frac{\nu}{c}, c\tau\right).$$

Hence, the AF of a space is "covariant" to TF shifts and scalings, convolutions and multiplications by chirp signals, and Fourier transforms. Note, however, that the covariance to TF shifts differs from that of the WD (cf. (2.33)).

11) Convolution and multiplication. A further important class of unitary transformations corresponds to a linear, time-invariant allpass filter. We here obtain

$$(\mathbf{H}x)(t) = \int_{t'} h(t-t')\, x(t')\, dt' \quad \text{with } |H(f)| \equiv 1$$

$$\Rightarrow \quad A_{\mathbf{H}\mathcal{X}}(\tau,\nu) = \int_{\tau'} A_h(\tau-\tau',\nu)\, A_{\mathcal{X}}(\tau',\nu)\, d\tau'. \tag{7.9}$$

Similarly, for the multiplication by a signal with unity envelope, we obtain

$$(\mathbf{H}x)(t) = m(t)\, x(t) \quad \text{with } |m(t)| \equiv 1$$

$$\Rightarrow \quad A_{\mathbf{H}\mathcal{X}}(\tau,\nu) = \int_{\nu'} A_m(\tau,\nu-\nu')\, A_{\mathcal{X}}(\tau,\nu')\, d\nu'. \tag{7.10}$$

Note that (7.7) and (7.8) are special cases of (7.9) and (7.10), respectively.

7.3   Some Signal Spaces

We now discuss the AF of the special signal spaces that were previously considered in Section 2.4 in the context of the WD.

1) Zero space. The AF of the trivial "zero space" $\mathcal{Z}$ whose only element is the zero signal $z(t) \equiv 0$ is identically zero,

$$A_{\mathcal{Z}}(\tau,\nu) \equiv 0\,.$$

2) One-dimensional space. The AF of a one-dimensional space $\mathcal{X}$ equals the AF of the single normalized basis signal $x_1(t)$,

$$A_{\mathcal{X}}(\tau,\nu) = A_{x_1}(\tau,\nu) \quad \text{for } N_{\mathcal{X}} = 1\,.$$

Thus, the AF of a (normalized) signal is a special case of the AF of a space.

3) Space of finite-energy signals. The AF of the infinite-dimensional "total space" $\mathcal{L}_2(\mathbb{R})$ of all finite-energy signals is a 2-D Dirac impulse at the origin of the (τ,ν)-plane,

$$A_{\mathcal{L}_2(\mathrm{IR})}(\tau,\nu) \;=\; \delta(\tau)\,\delta(\nu)\,.$$

Thus, the AF of $\mathcal{L}_2(\mathrm{IR})$ is perfectly concentrated at the origin of the (τ,ν)-plane.

4) Space of even signals. The AF of the space $\mathcal{E}$ comprising all even signals (i.e., signals satisfying $x(-t) = x(t)$) is

$$A_{\mathcal{E}}(\tau,\nu) \;=\; \frac{1}{2}\,\delta(\tau)\,\delta(\nu) \;+\; \frac{1}{4}\,.$$

The AF of any subspace of $\mathcal{E}$ is identical to the subspace's WD up to scalings:

$$\mathcal{X}\subseteq\mathcal{E} \quad\Rightarrow\quad A_{\mathcal{X}}(\tau,\nu) \;=\; \frac{1}{2}\,W_{\mathcal{X}}\!\left(\frac{\tau}{2},\frac{\nu}{2}\right).$$

5) Space of odd signals. The AF of the space $\mathcal{O}$ comprising all odd signals (i.e., signals satisfying $x(-t) = -x(t)$) is

$$A_{\mathcal{O}}(\tau,\nu) \;=\; \frac{1}{2}\,\delta(\tau)\,\delta(\nu) \;-\; \frac{1}{4}\,.$$

The AF of any subspace of $\mathcal{O}$ is related to the subspace's WD as follows:

$$\mathcal{X}\subseteq\mathcal{O} \quad\Rightarrow\quad A_{\mathcal{X}}(\tau,\nu) \;=\; -\frac{1}{2}\,W_{\mathcal{X}}\!\left(\frac{\tau}{2},\frac{\nu}{2}\right).$$

6) Space of time-limited signals. The AF of the space $\mathcal{T}[t_1,t_2]$ of all signals time-limited to a given interval $[t_1,t_2]$ is

$$A_{\mathcal{T}[t_1,t_2]}(\tau,\nu) \;=\; \delta(\tau)\,\tau_{12}\,\mathrm{sinc}(\pi\tau_{12}\,\nu)\,e^{-j2\pi t_{12}\nu}$$

with $t_{12} = \frac{t_1+t_2}{2}$, $\tau_{12} = t_2 - t_1$, and $\mathrm{sinc}(\beta) = \sin(\beta)/\beta$. Hence, the AF of $\mathcal{T}[t_1,t_2]$ is perfectly concentrated with respect to the time lag τ.

7) Space of causal signals. The AF of the space $\mathcal{C} = \mathcal{T}[0,\infty)$ of causal signals can be obtained by taking the 2-D Fourier transform of (2.41) as suggested by (7.5); this yields

$$A_{\mathcal{C}}(\tau,\nu) \;=\; \frac{1}{2}\,\delta(\tau)\left[\delta(\nu) + \underset{t\to\nu}{\mathcal{F}}\,\{\mathrm{sgn}(t)\}\right] \;=\; \frac{1}{2}\,\delta(\tau)\left[\delta(\nu) + \frac{1}{j\pi\nu}\right].$$

8) Space of band-limited signals. The AF of the space $\mathcal{F}[f_1,f_2]$ of all signals band-limited to a given frequency band $[f_1,f_2]$ is

$$A_{\mathcal{F}[f_1,f_2]}(\tau,\nu) \;=\; \delta(\nu)\,\nu_{12}\,\mathrm{sinc}(\pi\nu_{12}\tau)\,e^{j2\pi f_{12}\tau}$$

with $f_{12} = \frac{f_1+f_2}{2}$, $\nu_{12} = f_2 - f_1$.

9) Space of analytic signals. The AF of the space $\mathcal{A} = \mathcal{F}[0, \infty)$ of analytic signals is (cf. (2.42))

$$A_{\mathcal{A}}(\tau, \nu) = \frac{1}{2} \delta(\nu) \left[\delta(\tau) + \mathcal{F}^{-1}_{f \to \tau} \{ \mathrm{sgn}(f) \} \right] = \frac{1}{2} \delta(\nu) \left[\delta(\tau) - \frac{1}{j\pi\tau} \right] .$$

10) Hermite spaces. The AF of the kth Hermite function $h_k^{(T)}(t)$ as defined in (2.43) is given by [Wilcox, 1991, Klauder, 1960]

$$A_{h_k^{(T)}}(\tau, \nu) = \frac{(-1)^{k-1}}{2} W_{h_k^{(T)}}\left(\frac{\tau}{2}, \frac{\nu}{2} \right) = \tilde{v}_{k-1}\left(\left(\frac{\tau}{T}\right)^2 + (T\nu)^2 \right), \quad k = 1, 2, \dots$$

with (cf. (2.45))

$$\tilde{v}_n(\beta) = \frac{(-1)^n}{2} v_n\left(\frac{\beta}{4} \right) = L_n(\pi\beta)\, e^{-\frac{\pi\beta}{2}}, \qquad n = 0, 1, \dots,$$

where $L_n(\beta)$ is the Laguerre polynomial of order n (cf. (2.46)). The AF of the Hermite space $\mathcal{H}_N^{(T)} = \mathrm{span}\{ h_k^{(T)}(t) \}_{k=1}^N$ is then obtained as

$$A_{\mathcal{H}_N^{(T)}}(\tau, \nu) = \sum_{k=1}^{N} A_{h_k^{(T)}}(\tau, \nu) = \tilde{w}_N\left(\left(\frac{\tau}{T}\right)^2 + (T\nu)^2 \right)$$

with

$$\tilde{w}_N(\beta) = \sum_{n=0}^{N-1} \tilde{v}_n(\beta) = e^{-\frac{\pi\beta}{2}} \sum_{n=0}^{N-1} L_n(\pi\beta) .$$

We see that the AF of $\mathcal{H}_N^{(T)}$ shows "elliptical symmetry," i.e., its height is constant along any ellipse $(\tau/T)^2 + (T\nu)^2 = K^2$ in the (τ, ν)-plane. With $\sum_{n=0}^{N-1} L_n(\beta) = L_{N-1}^{(1)}(\beta)$ [Szegö, 1975], where $L_n^{(\alpha)}(\beta)$ denotes the generalized Laguerre polynomials [Abramowitz and Stegun, 1965, Szegö, 1975], we have $\tilde{w}_N(\beta) = e^{-\frac{\pi\beta}{2}} L_{N-1}^{(1)}(\pi\beta)$ and thus

$$A_{\mathcal{H}_N^{(T)}}(\tau, \nu) = \exp\left(-\frac{\pi}{2}\left[\left(\frac{\tau}{T}\right)^2 + (T\nu)^2 \right] \right) L_{N-1}^{(1)}\left(\pi\left[\left(\frac{\tau}{T}\right)^2 + (T\nu)^2 \right] \right) .$$

For $N = 1$, the Hermite space reduces to the single Gaussian function $h_1^{(T)}(t) = \sqrt{\sqrt{2}/T}\, e^{-\pi(t/T)^2}$ whose AF is a two-dimensional Gaussian,

$$A_{\mathcal{H}_1^{(T)}}(\tau, \nu) = A_{h_1^{(T)}}(\tau, \nu) = e^{-\frac{\pi}{2}\left[(\tau/T)^2 + (T\nu)^2 \right]} .$$

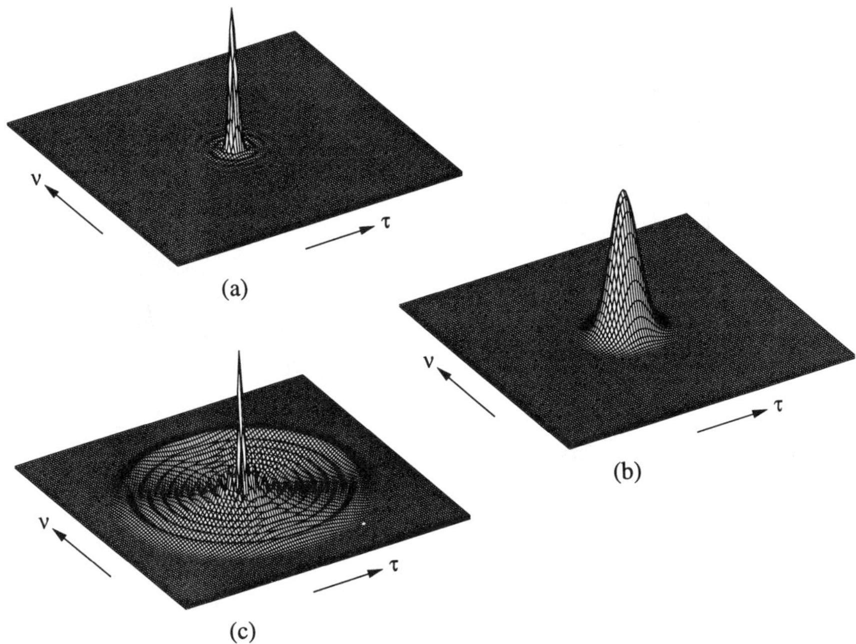

Figure 7.1.   AF of Hermite space and Hermite functions: (a) Squared magnitude of the AF of a 10-dimensional Hermite space, as compared to the squared magnitude of the AF of (b) the Hermite function of order 1 and (c) the Hermite function of order 10.

For $N \to \infty$, $\mathcal{H}_N^{(T)}$ becomes $\mathcal{L}_2(\mathbb{R})$, and thus

$$A_{\mathcal{H}_\infty^{(T)}}(\tau, \nu) \;=\; A_{\mathcal{L}_2(\mathbb{R})}(\tau, \nu) \;=\; \delta(\tau)\, \delta(\nu)\,.$$

Fig. 7.1 compares the AF of a Hermite space of dimension 10 with the AF of the first and the 10th Hermite function. We see that the peak of the space AF is sharper than that of the AF of $h_1^{(T)}(t)$ but less sharp than that of the AF of $h_{10}^{(T)}(t)$. On the other hand, the sidelobes in the space AF are considerably lower than those in the AF of $h_{10}^{(T)}(t)$.

11) Prolate spheroidal spaces. Due to the fact that the kth prolate spheroidal wave function $p_k^{(T,F)}(t)$ (see Section 2.4) is an even function for odd k and an odd function for even k, the AF of $p_k^{(T,F)}(t)$ is equal to the WD of $p_k^{(T,F)}(t)$ up to scalings and (for even k) a minus sign,

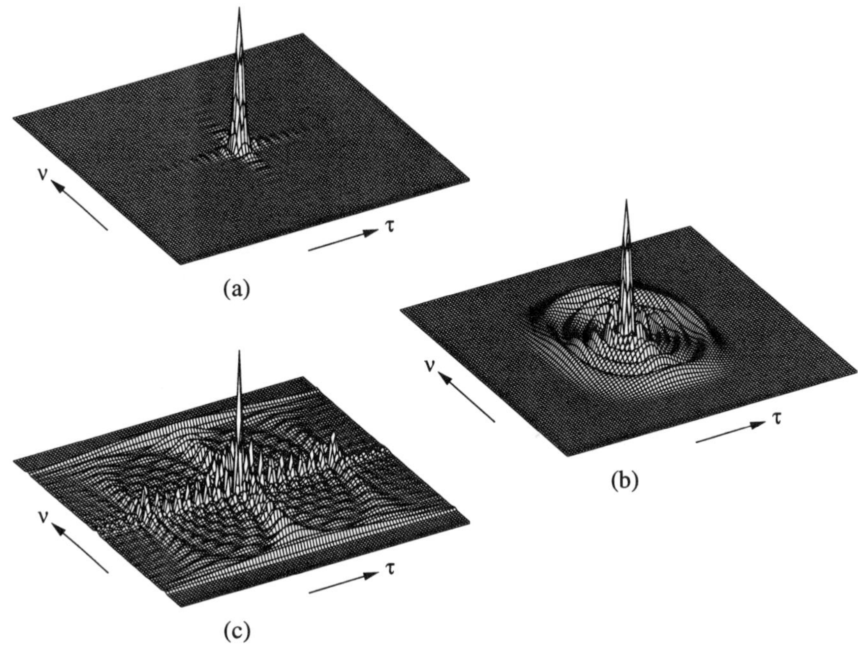

(a)

(b)

(c)

Figure 7.2.   AF of prolate spheroidal space and prolate spheroidal wave functions: (a) Squared magnitude of the AF of a 10-dimensional prolate spheroidal space, as compared to the squared magnitude of the AF of (b) the prolate spheroidal wave function of order 5 and (c) the prolate spheroidal wave function of order 10. In all cases, $TF = 10$.

$$A_{p_k^{(T,F)}}(\tau, \nu) = \frac{(-1)^{k-1}}{2} W_{p_k^{(T,F)}}\left(\frac{\tau}{2}, \frac{\nu}{2}\right), \qquad k = 1, 2, \ldots$$

Since $p_k^{(T,F)}(t)$ is band-limited to $[-F/2, F/2]$, the AF of $p_k^{(T,F)}(t)$ is zero for $|\nu| > F$. The same is true for the AF of the N-dimensional prolate spheroidal space $\mathcal{P}_N^{(T,F)} = \mathrm{span}\{p_k^{(T,F)}(t)\}_{k=1}^N$,

$$A_{\mathcal{P}_N^{(T,F)}}(\tau, \nu) = 0 \quad \text{for} \quad |\nu| > F.$$

Fig. 7.2 compares the AF of a prolate spheroidal space of dimension 10 with the AF of the 5th and 10th prolate spheroidal wave function. The sharpness of the peak of the space AF is similar to that of the AF of $p_5^{(T,F)}(t)$, while the sidelobes of the space AF are much lower than those of the signal AFs.

7.4 Interpretations

According to Section 2.5, the WD of a linear signal space can be interpreted as a TF distribution of the space's dimension (energy). The interpretation of the AF of a linear signal space is completely different. Two (related) interpretations of the AF will be described in the next two subsections.

7.4.1 Deterministic Interpretation

The AF expression (7.2),

$$A_{\mathcal{X}}(\tau,\nu) \;=\; \int_t P_{\mathcal{X}}\left(t+\frac{\tau}{2}, t-\frac{\tau}{2}\right) e^{-j2\pi\nu t}\, dt\,,$$

shows that the AF of a linear signal space is the *spreading function* [Bello, 1963, Sostrand, 1968, Kozek and Hlawatsch, 1991b, Kozek, 1992a, Kozek, 1997a, Kozek, 1997b] of the space's orthogonal projection operator $\mathbf{P}_{\mathcal{X}}$. This entails an interesting interpretation of the AF in terms of TF shifts.

Let $\mathbf{S}^{(\tau,\nu)}$ denote the elementary TF shift operator defined as

$$\left(\mathbf{S}^{(\tau,\nu)} s\right)(t) \;=\; s(t-\tau)\, e^{j2\pi\nu t}\,.$$

With (7.4), one can show that the orthogonal projection of a signal $s(t)$ onto the signal space $\mathcal{X}$ can be written as

$$(\mathbf{P}_{\mathcal{X}} s)(t) \;=\; \int_\tau \int_\nu A_{\mathcal{X}}(\tau,\nu)\, e^{-j\pi\tau\nu} \left(\mathbf{S}^{(\tau,\nu)} s\right)(t)\, d\tau\, d\nu\,.$$

This corresponds to the following expansion of the orthogonal projection operator $\mathbf{P}_{\mathcal{X}}$ into elementary TF shift operators $\mathbf{S}^{(\tau,\nu)}$:

$$\mathbf{P}_{\mathcal{X}} \;=\; \int_\tau \int_\nu A_{\mathcal{X}}(\tau,\nu)\, e^{-j\pi\tau\nu}\, \mathbf{S}^{(\tau,\nu)}\, d\tau\, d\nu\,. \tag{7.11}$$

The coefficient function in this expansion is, up to a phase factor $e^{-j\pi\tau\nu}$, the AF of $\mathcal{X}$. Thus, if $|A_{\mathcal{X}}(\tau,\nu)|$ is large about a TF lag point (τ_0,ν_0), then a significant part of the orthogonal projection $s_{\mathcal{X}}(t)$ is made up by the signal $s(t)$ TF shifted by (τ_0,ν_0). Since $|A_{\mathcal{X}}(-\tau,-\nu)| = |A_{\mathcal{X}}(\tau,\nu)|$, TF shifts by $(-\tau,-\nu)$ and by (τ,ν) are contained in $s_{\mathcal{X}}(t)$ to an equal extent.

This interpretation of the AF of a signal space as coefficient function in a TF shift expansion of the orthogonal projection operator shows that the AF of a signal space is closely related to the distinction between simple and sophisticated spaces introduced in Section 3.1.2. Let us reconsider the elementary sophisticated space previously discussed in Example 3.2.

Example 7.1. Consider the one-dimensional space $\mathcal{X}$ spanned by

$$x_1(t) = \frac{1}{\sqrt{2}}\left[y_1(t) + y_2(t)\right],$$

where $y_1(t)$ and $y_2(t)$ are two orthonormal signals with good TF concentration about the TF points (t_1, f_1) and (t_2, f_2), respectively. These two TF points are assumed to be sufficiently far apart. The AF of the space $\mathcal{X}$ is

$$
\begin{aligned}
A_{\mathcal{X}}(\tau, \nu) &= A_{x_1}(\tau, \nu) \\
&= \frac{1}{2}\left[A_{y_1}(\tau, \nu) + A_{y_2}(\tau, \nu) + A_{y_1, y_2}(\tau, \nu) + A^*_{y_1, y_2}(-\tau, -\nu)\right].
\end{aligned}
\tag{7.12}
$$

The "auto terms" $\frac{1}{2}A_{y_1}(\tau, \nu)$ and $\frac{1}{2}A_{y_2}(\tau, \nu)$ are concentrated about the origin of the (τ, ν)-plane. The "cross terms" $\frac{1}{2}A_{y_1, y_2}(\tau, \nu)$ and $\frac{1}{2}A^*_{y_1, y_2}(-\tau, -\nu)$, on the other hand, can be shown [Hlawatsch and Flandrin, 1997] to be located about the TF lag points (τ_0, ν_0) and $(-\tau_0, -\nu_0)$, respectively, where $\tau_0 = t_1 - t_2$ and $\nu_0 = f_1 - f_2$, i.e., $|\tau_0|$ and $|\nu_0|$ are the time and frequency distances, respectively, between the "signal points" (t_1, f_1) and (t_2, f_2). With (7.11), it follows that the orthogonal projection onto $\mathcal{X}$ must be expected to cause TF displacements by (τ_0, ν_0) and $(-\tau_0, -\nu_0)$. This is indeed readily verified. According to (3.3), the orthogonal projection of $y_1(t)$ onto $\mathcal{X}$ is

$$y_{1,\mathcal{X}}(t) = \frac{1}{2}\left[y_1(t) + y_2(t)\right].$$

While the input signal $y_1(t)$ is concentrated about the TF point (t_1, f_1), the projected signal $y_{1,\mathcal{X}}(t)$ has half of its energy concentrated about (t_2, f_2). Thus, signal energy has been shifted from (t_1, f_1) to (t_2, f_2), which is a TF shift by $(-\tau_0, -\nu_0)$. Projecting $y_2(t)$ instead of $y_1(t)$ shows that a TF shift by (τ_0, ν_0) is possible too. $\qquad\square$

In the elementary example discussed above, the AF of $\mathcal{X}$ contained significant "cross terms" that were concentrated about the two TF lag points (τ_0, ν_0) and $(-\tau_0, -\nu_0)$ located far away from the origin of the (τ, ν)-plane. These "off-origin" AF cross terms correspond (via the 2-D Fourier transform, see (7.5)) to oscillatory WD cross terms that are characteristic of sophisticated spaces. Indeed, strong TF shift effects of the orthogonal projection operator, strong oscillatory cross terms in the WD, and strong off-origin cross terms in the AF are three *equivalent* indications of a space's sophistication. Hence, a space is simple if its AF is well concentrated about the origin of the (τ, ν)-plane and sophisticated if its AF contains significant components far away from the origin. We conclude that the AF of a signal space is a TF space representation particularly suited for assessing and analyzing the sophistication of a space.

7.4.2 Stochastic Interpretation

A second interpretation of the AF of a linear signal space is conceptually related to the interpretation discussed above. Let $w(t)$ be wide-sense stationary, zero-mean white noise with normalized power spectral density. The AF of $w(t)$ is a random function of τ and ν. The expectation of the AF is easily shown to be

$$\bar{A}_w(\tau,\nu) \;\triangleq\; \mathrm{E}\{A_w(\tau,\nu)\} \;=\; \delta(\tau)\,\delta(\nu)\,. \qquad (7.13)$$

According to the general principle of defining quadratic space representations (see Section 2.2), the AF of a linear signal space $\mathcal{X}$ is the expected AF of the orthogonal projection $w_{\mathcal{X}}(t) = \big(\mathbf{P}_{\mathcal{X}} w\big)(t)$ of $w(t)$,

$$A_{\mathcal{X}}(\tau,\nu) \;=\; \bar{A}_{w_{\mathcal{X}}}(\tau,\nu)\,. \qquad (7.14)$$

In order to interpret this identity, we shall now show that the expected AF of a (generally) nonstationary random process $x(t)$ can be interpreted as a "TF correlation function" [Kozek et al., 1994, Kozek, 1996, Kozek, 1997a]. Let us assume that the process $x(t)$ has signal components around two TF points (t_1, f_1) and (t_2, f_2). We would like to measure the statistical correlation between these two TF points in a similar way as the autocorrelation function $R_x(t_1, t_2) = \mathrm{E}\{x(t_1)\, x^*(t_2)\}$ measures the statistical correlation between two time points t_1 and t_2. An intuitively reasonable measure for the statistical correlation between two TF points (t_1, f_1) and (t_2, f_2) is

$$C_x(t_1, f_1; t_2, f_2) \;\triangleq\; \mathrm{E}\big\{\langle x, h_1\rangle\, \langle x, h_2\rangle^*\big\}\,,$$

where $h_1(t)$ and $h_2(t)$ are two normalized, deterministic "test signals" that are TF localized about (t_1, f_1) and (t_2, f_2), respectively (see part (a) of **Fig. 7.3**). The inner product $\langle x, h_i\rangle = \int x(t)\, h_i^*(t)\, dt$ $(i = 1, 2)$ measures the "content of $x(t)$ about the TF point (t_i, f_i)." With Moyal's formula [Wilcox, 1991] $\langle x, h_1\rangle\, \langle x, h_2\rangle^* = \langle A_x, A_{h_1,h_2}\rangle$, where $A_{h_1,h_2}(\tau,\nu)$ is the cross-AF of the test signals $h_1(t)$ and $h_2(t)$, it follows that the TF correlation $C_x(t_1, f_1; t_2, f_2)$ can be derived from the expected AF $\bar{A}_x(\tau,\nu)$ as

$$C_x(t_1, f_1; t_2, f_2) \;=\; \langle \bar{A}_x, A_{h_1,h_2}\rangle \;=\; \int_\tau \int_\nu \bar{A}_x(\tau,\nu)\, A_{h_1,h_2}^*(\tau,\nu)\, d\tau\, d\nu\,. \qquad (7.15)$$

For signals $h_1(t)$ and $h_2(t)$ TF localized about (t_1, f_1) and (t_2, f_2), respectively, $A_{h_1,h_2}(\tau,\nu)$ is known to be located about the TF lag point (τ_0, ν_0), where $\tau_0 = t_1 - t_2$ and $\nu_0 = f_1 - f_2$ [Hlawatsch and Flandrin, 1997]. If (τ_0, ν_0) is well outside the effective support of the expected AF $\bar{A}_x(\tau,\nu)$ (as shown in Fig. 7.3(b)), then $\langle \bar{A}_x, A_{h_1,h_2}\rangle = 0$ and, due to (7.15), also $C_x(t_1, f_1; t_2, f_2) = 0$.

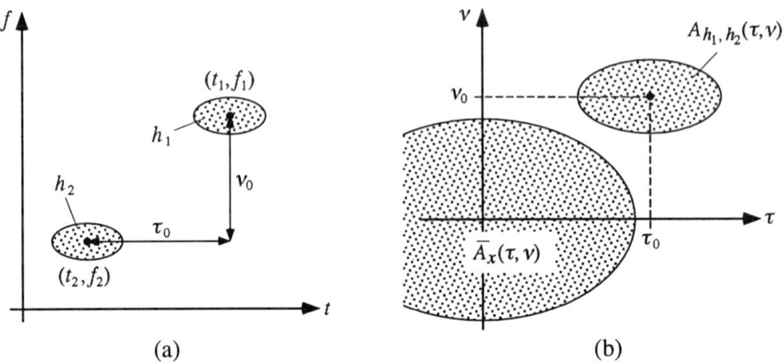

Figure 7.3. TF correlation interpretation of the expected AF: (a) TF plane, (b) TF lag plane.

We have thus shown the following result: If the expected AF is zero about a given "TF lag point" (τ_0, ν_0), then any two TF points (t_1, f_1) and (t_2, f_2) with $t_1 - t_2 = \tau_0$ and $f_1 - f_2 = \nu_0$ are uncorrelated.

Conversely, if the TF lag point (τ_0, ν_0) is inside the effective support of $\bar{A}_x(\tau, \nu)$ so that $\bar{A}_x(\tau, \nu)$ and $A_{h_1, h_2}(\tau, \nu)$ overlap, this does not necessarily imply that the TF points (t_1, f_1) and (t_2, f_2) are correlated: since both $\bar{A}_x(\tau, \nu)$ and $A_{h_1, h_2}(\tau, \nu)$ are typically oscillatory functions, $\langle \bar{A}_x, A_{h_1, h_2} \rangle$ may still be zero. Thus, the expected AF $\bar{A}_x(\tau, \nu)$ indicates the *potential correlation* between TF points separated by a time lag τ and a frequency lag ν.

After this discussion of the expected AF of a random process, we now return to the AF of a signal space $\mathcal{X}$. According to (7.14), $A_{\mathcal{X}}(\tau, \nu)$ is the expected AF of $w_{\mathcal{X}}(t) = (\mathbf{P}_{\mathcal{X}} w)(t)$. The result $\bar{A}_w(\tau, \nu) = \delta(\tau)\,\delta(\nu)$ in (7.13) shows that in the case of stationary white noise $w(t)$ different TF points are uncorrelated. Projecting $w(t)$ onto $\mathcal{X}$ will generally introduce correlation between different TF points; this correlation is indicated (in the sense discussed above) by the AF of $\mathcal{X}$. In particular, if the orthogonal projection causes (part of) a component of $w(t)$ that is located about a TF point (t_1, f_1) to be shifted to another TF point (t_2, f_2), then the two TF points (t_1, f_1) and (t_2, f_2), originally uncorrelated in $w(t)$, will be correlated in $w_{\mathcal{X}}(t) = (\mathbf{P}_{\mathcal{X}} w)(t)$. This shows that nonzero TF correlation in the process $w_{\mathcal{X}}(t)$, indicated by nonzero values of $A_{\mathcal{X}}(\tau, \nu) = \bar{A}_{w_{\mathcal{X}}}(\tau, \nu)$ away from the origin of the (τ, ν)-plane, can be attributed to TF shifts caused by the orthogonal projection operator $\mathbf{P}_{\mathcal{X}}$. Thus, we have established a conceptual link to the "deterministic" interpretation discussed in Section 7.4.1.

Example 7.2. Let us consider the TF correlation of the process $w_{\mathcal{X}}(t) = (\mathbf{P}_{\mathcal{X}} w)(t)$ where $\mathcal{X}$ is the one-dimensional space from Example 7.1. We recall that this space is spanned by the single basis signal $x_1(t) = \frac{1}{\sqrt{2}}[y_1(t) + y_2(t)]$, where $y_1(t)$ and $y_2(t)$ are two orthonormal signals with good TF concentration about the TF points (t_1, f_1) and (t_2, f_2), respectively. With (7.12), the expected AF of $w_{\mathcal{X}}(t)$ is obtained as

$$
\begin{aligned}
\bar{A}_{w_{\mathcal{X}}}(\tau, \nu) &= A_{\mathcal{X}}(\tau, \nu) \\
&= \frac{1}{2}\left[A_{y_1}(\tau, \nu) + A_{y_2}(\tau, \nu) + A_{y_1, y_2}(\tau, \nu) + A_{y_1, y_2}^*(-\tau, -\nu)\right].
\end{aligned}
$$

Since the "cross terms" $\frac{1}{2} A_{y_1, y_2}(\tau, \nu)$ and $\frac{1}{2} A_{y_1, y_2}^*(-\tau, -\nu)$ are located about the TF lag points $(\pm\tau_0, \pm\nu_0)$ with $\tau_0 = t_1 - t_2$ and $\nu_0 = f_1 - f_2$, the process $w_{\mathcal{X}}(t)$ has potentially nonzero TF correlation about any two TF points separated by $(\pm\tau_0, \pm\nu_0)$. This TF correlation can be attributed to the TF shifts caused by the projection operator $\mathbf{P}_{\mathcal{X}}$ as discussed in Example 7.1. $\qquad\Box$

7.5 Extensions

We next introduce and discuss the cross-AF of two spaces and a discrete-time AF version.

7.5.1 *Cross-Ambiguity Function*

The definition of the cross-AF (CAF) $A_{\mathcal{X}, \mathcal{Y}}(\tau, \nu)$ of two signal spaces $\mathcal{X}$ and $\mathcal{Y}$ is analogous to that of the cross-WD (see Section 2.6.1). The CAF can be expressed in terms of the spaces' orthogonal projection operators $\mathbf{P}_{\mathcal{X}}$ and $\mathbf{P}_{\mathcal{Y}}$ or orthonormal bases $\{x_k(t)\}_{k=1}^{N_{\mathcal{X}}}$ and $\{y_l(t)\}_{l=1}^{N_{\mathcal{Y}}}$ as (cf. (7.2) and (7.3))

$$
\begin{aligned}
A_{\mathcal{X}, \mathcal{Y}}(\tau, \nu) &= \int_t (\mathbf{P}_{\mathcal{X}} \mathbf{P}_{\mathcal{Y}}^+)\left(t + \frac{\tau}{2}, t - \frac{\tau}{2}\right) e^{-j2\pi\nu t}\, dt \\
&= \sum_{k=1}^{N_{\mathcal{X}}} \sum_{l=1}^{N_{\mathcal{Y}}} \langle y_l, x_k\rangle\, A_{x_k, y_l}(\tau, \nu)
\end{aligned}
$$

where $(\mathbf{P}_{\mathcal{X}} \mathbf{P}_{\mathcal{Y}}^+)(t, t')$ denotes the kernel of the composite operator $\mathbf{P}_{\mathcal{X}} \mathbf{P}_{\mathcal{Y}}^+$. The last expression can also be written as

$$
A_{\mathcal{X}, \mathcal{Y}}(\tau, \nu) = \sum_{k=1}^{N_{\mathcal{X}}} A_{x_k, x_{k}, \mathcal{Y}}(\tau, \nu) = \sum_{l=1}^{N_{\mathcal{Y}}} A_{y_l, \mathcal{X}, y_l}(\tau, \nu),
$$

where $x_{k, \mathcal{Y}}(t)$ and $y_{l, \mathcal{X}}(t)$ are the projections of $x_k(t)$ onto $\mathcal{Y}$ and of $y_l(t)$ onto $\mathcal{X}$, respectively. The CAF satisfies

$$
A_{\mathcal{Y}, \mathcal{X}}(\tau, \nu) = A_{\mathcal{X}, \mathcal{Y}}^*(-\tau, -\nu) \qquad \text{and} \qquad A_{\mathcal{X}, \mathcal{X}}(\tau, \nu) = A_{\mathcal{X}}(\tau, \nu),
$$

and it is essentially the 2-D Fourier transform of the CWD,

$$A_{\mathcal{X},\mathcal{Y}}(\tau,\nu) = \int_t \int_f W_{\mathcal{X},\mathcal{Y}}(t,f)\, e^{-j2\pi(\nu t - \tau f)}\, dt\, df \,.$$

Besides many properties that extend properties of the auto-AF (cf. Section 7.2), two remarkable properties of the CAF are

$$A_{\mathcal{X},\mathcal{Y}}(\tau,\nu) \equiv 0 \quad\Longleftrightarrow\quad \mathcal{X} \perp \mathcal{Y}$$

and

$$A_{\mathcal{X},\mathcal{Y}}(\tau,\nu) = A_{\mathcal{X}}(\tau,\nu) \quad\Longleftrightarrow\quad \mathcal{X} \subseteq \mathcal{Y} \,.$$

It should be noted that the CAF of two spaces is only partly analogous to the CAF of two signals. In particular, while the CAF of two signals occurs in the "quadratic superposition law"

$$A_{x+y}(\tau,\nu) = A_x(\tau,\nu) + A_y(\tau,\nu) + A_{x,y}(\tau,\nu) + A_{x,y}^*(-\tau,-\nu) \,,$$

a similar expression does not generally exist for the AF $A_{\mathcal{X}+\mathcal{Y}}(\tau,\nu)$ of the sum of two spaces. Furthermore, the CAF of two one-dimensional spaces $\mathcal{X}$, $\mathcal{Y}$ is generally *not* equal to the CAF of the (normalized) basis signals $x_1(t)$, $y_1(t)$. These remarks parallel the remarks on the CWD given in Section 2.6.1.

7.5.2 Discrete-Time Ambiguity Function

The discrete-time AF of a discrete-time signal $x(n)$ can be defined as

$$A_x^{(D)}(m,\zeta) = \sum_n x(n+m)\, x^*(n-m)\, e^{-j2\pi\zeta n} \,,$$

where m is a discrete time lag and ζ is a normalized frequency lag. Let $\mathcal{X}$ be a linear space of discrete-time signals $x(n)$ which is characterized by its orthogonal projection operator $\mathbf{P}_{\mathcal{X}}$ with kernel $P_{\mathcal{X}}(n,n')$ or by an orthonormal basis $\{x_k(n)\}_{k=1}^{N_{\mathcal{X}}}$. We define the discrete-time AF of the space $\mathcal{X}$ as

$$A_{\mathcal{X}}^{(D)}(m,\zeta) = \sum_n P_{\mathcal{X}}(n+m,\, n-m)\, e^{-j2\pi\zeta n} \tag{7.16}$$

$$= \sum_{k=1}^{N_{\mathcal{X}}} A_{x_k}^{(D)}(m,\zeta) \tag{7.17}$$

(cf. (7.2) and (7.3)). The discrete-time AF is periodic, with respect to the variable ζ, with period 1. It is related to the discrete-time WD as

$$A_{\mathcal{X}}^{(D)}(m,\zeta) = \sum_n \int_0^{1/2} W_{\mathcal{X}}^{(D)}(n,\theta)\, e^{-j2\pi(\zeta n - 2m\theta)}\, d\theta \,.$$

Let us now assume that the discrete-time space $\mathcal{X}$ is a sampled version of a band-limited continuous-time space $\mathcal{X}_c$ as defined in Section 2.6.2,

$$\mathcal{X} \stackrel{\triangle}{=} \left\{ x(n) \; : \; x(n) = x_c(nT) \text{ with } x_c(t) \in \mathcal{X}_c \right\}.$$

We assume that the conventional sampling condition is satisfied, i.e., $T \leq 1/F$ where F is the (total) bandwidth of $\mathcal{X}$. In this case, it can be shown that the discrete-time AF of $\mathcal{X}$ and the continuous-time AF of $\mathcal{X}_c$ are related as

$$A_{\mathcal{X}}^{(D)}(m, \zeta) \; = \; \sum_{l=-\infty}^{\infty} A_{\mathcal{X}_c}\left(2mT, \frac{\zeta - l}{T} \right). \tag{7.18}$$

We see that the continuous-time AF is periodized, with respect to ν, with frequency lag period $1/T$. Since the continuous-time AF $A_{\mathcal{X}_c}(\tau, \nu)$ of a signal space $\mathcal{X}_c$ with total bandwidth F may assume nonzero values for $|\nu| \leq F$, and since by assumption $1/T \geq F$, it follows from (7.18) that the discrete-time AF may contain aliasing. Furthermore, we see that the continuous-time AF is sampled, with respect to τ, with sampling period $2T$ rather than T, which may cause an information loss. The information loss caused by the aliasing and subsampling will however be avoided if the total bandwidth of the signal space $\mathcal{X}_c$ is not larger than half the sampling rate, i.e., $F \leq f_s/2$ with $f_s = 1/T$. In this case, the discrete-time space $\mathcal{X}$ will be a subspace of some halfband space $\mathcal{H}^{(\theta_0)}$ (cf. Section 2.6.2), and relation (7.18) simplifies to

$$A_{\mathcal{X}}^{(D)}(m, \zeta) \; = \; A_{\mathcal{X}_c}\left(2mT, \frac{\zeta}{T} \right) \qquad \text{for } |\zeta| < 1/2.$$

We see that aliasing is avoided; furthermore, it can be shown that the sampling period $2T$ is now adequate. Note that $F \leq f_s/2$ is identical to the condition that guarantees the discrete-time WD to be non-aliased (cf. Section 2.6.2).

According to (7.16), a version of $A_{\mathcal{X}}^{(D)}(m, \zeta)$ that is discretized with respect to ζ can be computed by applying the FFT to the projection operator kernel $P_{\mathcal{X}}(n, n') = \sum_{k=1}^{N_{\mathcal{X}}} x_k(n)\, x_k^*(n')$. This is more efficient than adding the AFs of all basis signals $x_k(n)$ as suggested by (7.17).

8 RANGE-DOPPLER ESTIMATION

Perhaps the most important application of the AF of a signal is the radar/sonar problem of jointly estimating the range and radial velocity (Doppler shift) of a slowly fluctuating point target. The AF is important in this context since (i) the maximum-likelihood estimator for range and Doppler shift uses the cross-AF of the transmitted and received signals (or "pulses"), and (ii) the performance of this estimator is characterized by the AF of the transmitted pulse.

In this chapter, we show that a similar application exists for the *AF of a linear signal space* which has been introduced in Chapter 7. This application involves a generalized maximum-likelihood estimator that is based on a set of transmitted pulses, rather than a single pulse. Under appropriate conditions, the AF of the signal space spanned by the transmitted pulses characterizes the performance of this "multipulse estimator" in exactly the same manner as the AF of a single pulse characterizes the performance of the conventional single-pulse estimator. We also show that a fundamental performance limitation of single-pulse estimators known as "radar uncertainty principle" is relaxed in the case of the multipulse estimator.

This chapter is organized as follows. Section 8.1 briefly reviews the classical single-pulse estimator. In Section 8.2, we introduce the new multipulse estima-

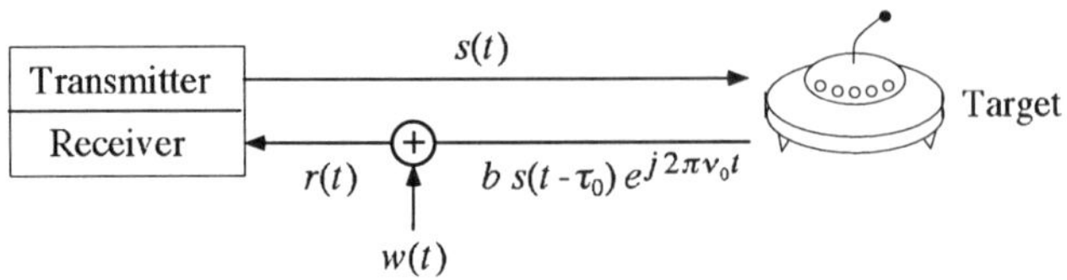

Figure 8.1.   A radar scenario.

tor and study its performance. It is shown that a global performance measure is optimized for orthogonal, equal-energy pulses. This globally optimum case is further considered in Section 8.3, where we show how the AF of the pulse space characterizes the estimator's performance.

8.1   The Single-Pulse Estimator and Its Performance

We consider the classical radar/sonar problem of estimating the range and radial velocity of a slowly fluctuating point target [Rihaczek, 1969, Van Trees, 1992, Woodward, 1953, Cook and Bernfeld, 1993, Skolnik, 1980], as illustrated schematically in **Fig. 8.1**. The active radar/sonar system transmits a narrowband signal or pulse[1] $s(t)$ with energy E_s. The transmitted signal is reflected by the target; the time delay and Doppler frequency shift of the received (reflected) signal are proportional to the target's range (i.e., distance from the transmitter/receiver) and radial velocity, respectively, and are to be estimated. The received signal can be modeled as [Van Trees, 1992]

$$r(t) = b\,s(t-\tau_0)\,e^{j2\pi\nu_0 t} + w(t).\tag{8.1}$$

Here, b is a zero-mean, circular complex, Gaussian random variable with variance $\mathrm{E}\{|b|^2\}=2\sigma_b^2$, $w(t)$ is circular complex, white Gaussian noise (statistically independent of b) with power spectral density η, τ_0 is a time lag (proportional to the target's range), and ν_0 is a Doppler frequency shift (approximately proportional to the target's radial velocity relative to the transmitter/receiver).

It is well known [Van Trees, 1992] that the maximum likelihood (ML) estimator of the parameter pair (τ_0, ν_0) is given by

$$(\hat{\tau}, \hat{\nu})_{\mathrm{ML}} = \arg\max_{\tau, \nu} |A_{r,s}(\tau, \nu)|^2,\tag{8.2}$$

where $A_{r,s}(\tau, \nu)$ is the cross-AF of the received signal $r(t)$ and the transmitted signal $s(t)$. The squared magnitude of the AF is known as the *ambiguity surface*

---

[1]The signals considered here are the complex baseband versions (complex envelopes) of the real bandpass waveforms actually transmitted or received. "Narrowband" means that the bandwidth of the transmitted signal is small compared to the carrier frequency. This assumption is necessary for approximating the Doppler effect by a frequency shift of the signal.

(AS). Eq. (8.2) shows that the ML estimate of the time lag τ and the Doppler frequency shift ν is the point in the time-lag/frequency-lag plane where the cross-AS of $s(t)$ and $r(t)$ has maximum height.

8.1.1 Local Accuracy

In what follows, it will be convenient to introduce the *normalized auto-AF*

$$\tilde{A}_s(\tau,\nu) \;\triangleq\; \frac{A_s(\tau,\nu)}{A_s(0,0)} \;=\; \frac{A_s(\tau,\nu)}{E_s} \,,$$

for which

$$\left|\tilde{A}_s(\tau,\nu)\right| \;\leq\; \tilde{A}_s(0,0) \;=\; 1 \,.$$

Two aspects of the performance of the single-pulse ML estimator are of particular interest. First, for high signal-to-noise ratio (SNR) the estimator's "local accuracy" can be approximately characterized by the Cramér-Rao lower bound (CRLB) which bounds the variance of any unbiased estimator[2] [Kay, 1993, Sorenson, 1980]. Under the simplifying assumption that the transmitted signal $s(t)$ is real-valued or conjugate even (i.e., $s(-t) = s^*(t)$), the CRLBs for the parameters τ and ν are given by [Van Trees, 1992]

$$\text{var}\{\hat{\tau}\} \;\geq\; \frac{-C}{\frac{d^2}{d\tau^2}|\tilde{A}_s(\tau,0)|^2\big|_{\tau=0}} \,, \qquad \text{var}\{\hat{\nu}\} \;\geq\; \frac{-C}{\frac{d^2}{d\nu^2}|\tilde{A}_s(0,\nu)|^2\big|_{\nu=0}} \,, \qquad (8.3)$$

with the SNR-dependent factor

$$C = \frac{\eta}{\overline{E}_r}\left(1 + \frac{\eta}{\overline{E}_r}\right) ,$$

where

$$\overline{E}_r \;\triangleq\; 2\sigma_b^2 E_s$$

is the mean received energy. The CRLBs depend on the SNR $\overline{E}_r/\eta$ through C and on the transmitted pulse $s(t)$ through the second partial derivatives (curvature) of $|\tilde{A}_s(\tau,\nu)|^2$ at the origin. The CRLBs are low (corresponding to potentially accurate estimates) if the peak of $|\tilde{A}_s(\tau,\nu)|^2$ at the origin of the (τ,ν)-plane (i.e., at the position of the maximum) is very sharp.

The CRLBs can be given an interesting interpretation [Van Trees, 1992]. It can be shown that

$$\frac{d^2}{d\tau^2}|\tilde{A}_s(\tau,0)|^2\bigg|_{\tau=0} = -2(2\pi)^2\tilde{F}_s^2 \,, \qquad \frac{d^2}{d\nu^2}|\tilde{A}_s(0,\nu)|^2\bigg|_{\nu=0} = -2(2\pi)^2\tilde{T}_s^2 \,,$$

---

[2]We note that ML estimators are not unbiased in general. However, they are asymptotically unbiased [Kay, 1993, Sorenson, 1980].

where $\tilde{T}_s$ and $\tilde{F}_s$ are the centered root-mean-square duration and bandwidth, respectively, of the transmitted pulse $s(t)$. These quantities are defined as

$$\tilde{T}_s^2 \stackrel{\triangle}{=} T_s^2 - t_s^2, \qquad \tilde{F}_s^2 \stackrel{\triangle}{=} F_s^2 - f_s^2 \tag{8.4}$$

with

$$T_s^2 = \frac{1}{E_s} \int_t t^2 \, |s(t)|^2 \, dt \,, \qquad t_s = \frac{1}{E_s} \int_t t \, |s(t)|^2 \, dt \tag{8.5}$$

and

$$F_s^2 = \frac{1}{E_s} \int_f f^2 \, |S(f)|^2 \, df \,, \qquad f_s = \frac{1}{E_s} \int_f f \, |S(f)|^2 \, df \,, \tag{8.6}$$

where $S(f)$ is the Fourier transform of $s(t)$. Hence, the CRLBs become

$$\mathrm{var}\{\hat{\tau}\} \ge \frac{C}{2\,(2\pi)^2} \frac{1}{\tilde{F}_s^2} \,, \qquad \mathrm{var}\{\hat{\nu}\} \ge \frac{C}{2\,(2\pi)^2} \frac{1}{\tilde{T}_s^2} \,, \tag{8.7}$$

which shows that the range and Doppler estimation accuracies may both be good only if $s(t)$ has simultaneously large bandwidth and large duration.

8.1.2   Global Accuracy

A second performance aspect is important for low SNR and/or in the presence of clutter (reverberation). Here, the maximum of $|A_{r,s}(\tau,\nu)|^2$ (i.e., the ML estimate) may occur at a totally wrong position ("outlier"). A simple measure of the estimator's "outlier immunity" or "global accuracy" is the expectation of $|A_{r,s}(\tau,\nu)|^2$. It is convenient to normalize by the value of $\mathrm{E}\{|A_{r,s}(\tau,\nu)|^2\}$ obtained at the true parameter point (τ_0,ν_0) in the absence of noise[3]:

$$\kappa_s(\tau,\nu) \stackrel{\triangle}{=} \frac{\mathrm{E}\{|A_{r,s}(\tau,\nu)|^2\}}{\mathrm{E}\{|A_{r,s}(\tau_0,\nu_0)|^2\}\big|_{w(t)\equiv 0}} \,.$$

Using (8.1) and the statistical independence of b and $w(t)$, it can be shown that $\kappa_s(\tau,\nu)$ can be expressed in terms of the normalized auto-AS as

$$\kappa_s(\tau,\nu) = \left|\tilde{A}_s(\tau-\tau_0,\nu-\nu_0)\right|^2 + \frac{\eta}{\overline{E}_r} \,. \tag{8.8}$$

For good global accuracy, $\kappa_s(\tau,\nu)$ must be small away from the true parameter pair (τ_0,ν_0). According to (8.8), $\kappa_s(\tau,\nu)$ consists of a signal-dependent term $|\tilde{A}_s(\tau-\tau_0,\nu-\nu_0)|^2$ peaked at the true parameter value (τ_0,ν_0), and a constant "SNR floor" $\eta/\overline{E}_r$. Thus, good global accuracy requires that $|\tilde{A}_s(\tau,\nu)|^2$ be small away from its peak at the origin of the (τ,ν)-plane.

---

[3]In the absence of noise, the only random quantity is the random factor b in (8.1); hence, the expectation is here with respect to b only.

8.1.3   The Radar Uncertainty Principle

We have seen that for good local as well as global accuracy, the normalized AS of the transmitted pulse, $|\tilde{A}_s(\tau, \nu)|^2$, should have a "thumbtack" shape, i.e., a sharp peak at the origin of the (τ, ν)-plane and very small values away from the origin. Since the peak height of $|\tilde{A}_s(\tau, \nu)|^2$ is fixed for all $s(t)$, $|\tilde{A}_s(0, 0)|^2 = 1$, a global measure for the thumbtack shape is the volume under $|\tilde{A}_s(\tau, \nu)|^2$,

$$V_s \triangleq \int_\tau \int_\nu \left|\tilde{A}_s(\tau, \nu)\right|^2 d\tau\, d\nu .$$

A better thumbtack shape requires a smaller volume V_s. Unfortunately, it can be shown that the volume always equals the peak height, i.e., there is

$$V_s = \left|\tilde{A}_s(0, 0)\right|^2 = 1 \qquad \text{for any } s(t) .$$

Thus, if we narrow the AS peak in order to increase the estimator's local accuracy as motivated by the CRLBs, the volume removed will reappear somewhere away from the AS peak and will thus reduce the estimator's global accuracy. This "radar uncertainty principle" limits the performance of the single-pulse estimator [Van Trees, 1992, Price and Hofstetter, 1965].

8.2   The Multipulse Estimator and Its Performance

Since the performance of the single-pulse estimator is restricted by the radar uncertainty principle, it is natural to ask whether an improvement can be obtained by using several pulses instead of a single pulse. A ML "multipulse estimator" [Hlawatsch and Edelson, 1992] will therefore be proposed and analyzed in this section. A relation to the AF of a linear signal space will be established in Section 8.3.

8.2.1   The Multipulse Estimator

Let us consider a radar/sonar system transmitting N pulses $s_k(t)$, $k = 1, ..., N$, with energies E_{s_k}. The overall transmitted energy, i.e., the energy of the transmitted pulse set $\{s_k(t)\}_{k=1}^N$, is

$$E_{s,N} \triangleq \sum_{k=1}^N E_{s_k} .$$

With the idealizing assumption that the pulses are transmitted and received simultaneously and independently of each other[4], the received signals can be

---

[4]In practice, this can be implemented only in an approximate manner using time multiplexing and/or frequency multiplexing (see the comments at the end of this subsection).

modeled as

$$r_k(t) \;=\; b\, s_k(t-\tau_0)\, e^{j2\pi\nu_0 t} + w_k(t)\,, \qquad k = 1, ..., N\,. \tag{8.9}$$

Here, the Gaussian random variable b with $\mathrm{E}\{|b|^2\} = 2\sigma_b^2$ is assumed to take on the same value for all k, and the white Gaussian noise processes $w_k(t)$ with power spectral density η are assumed to be statistically independent of each other and of b. The next theorem, whose proof can be found in Appendix 8.A, characterizes the ML estimator of range τ_0 and Doppler shift ν_0.

Theorem 8.1 (Multipulse ML Range-Doppler Estimator) *Given the N received signals $r_k(t)$ in (8.9) and the assumptions stated above, the ML estimator of range τ_0 and Doppler shift ν_0 is*

$$(\hat{\tau}, \hat{\nu})_{\mathrm{ML}} \;=\; \arg\max_{\tau,\nu} \left| \sum_{k=1}^{N} A_{r_k, s_k}(\tau, \nu) \right|^2 . \tag{8.10}$$

Comparing with the single-pulse estimator (8.2), we see that the cross-AF of $r(t)$ and $s(t)$ is replaced by the sum of the cross-AFs of $r_k(t)$ and $s_k(t)$.

Before discussing the theoretical performance of the multipulse estimator, we note that our assumptions will not be exactly satisfied in practical situations. In particular, it is impossible to transmit and receive N pulses "in parallel." Instead, we have to use time multiplexing (i.e., the pulses are transmitted one after the other) or frequency multiplexing (the pulses are transmitted during the same time interval, but about different carrier frequencies), or a combination of both. Let us briefly discuss potential problems of the time multiplex realization; analogous problems exist when frequency multiplexing is used.

First, since (for nonzero radial velocity) the target moves with respect to the radar/sonar system, the roundtrip delay will be different for each pulse. (In contrast, we assumed that all pulses are delayed by the same time τ_0.) This "range walk" has to be compensated, i.e., the cross-AFs $A_{r_k, s_k}(\tau, \nu)$ have to be aligned with respect to the τ parameter before carrying out the summation in (8.10). An exact compensation would require exact knowledge of the radial velocity; however, this is one of the parameters to be estimated. Approximate schemes for range-walk compensation can be devised but their discussion is beyond the scope of this work. A second assumption that may not be satisfied in practice is that of "pulse-to-pulse coherence," i.e., the assumption that the complex amplitude b in (8.9) is the same for all received pulses $r_k(t)$. With a time multiplex implementation, this assumption will be violated if the target fluctuates significantly during the time in which it is illuminated by the pulses. In the case of frequency multiplexing, b may depend on the carrier frequencies about which the individual pulses are located.

8.2.2 Local Accuracy

For a characterization of the performance of the multipulse ML estimator, it will be convenient to define the *normalized AF of the transmitted pulse set* $\{s_k(t)\}_{k=1}^N$ as

$$\tilde{A}_{s,N}(\tau,\nu) \triangleq \frac{1}{E_{s,N}} \sum_{k=1}^N A_{s_k}(\tau,\nu) , \qquad (8.11)$$

which is the sum of the individual auto-AFs $A_{s_k}(\tau,\nu)$ normalized by the peak value, i.e., the overall transmitted energy $E_{s,N} = \sum_{k=1}^N E_{s_k} = \sum_{k=1}^N A_{s_k}(0,0)$. The peak height of $\tilde{A}_{s,N}(\tau,\nu)$ again occurs at the origin of the (τ,ν)-plane,

$$\left|\tilde{A}_{s,N}(\tau,\nu)\right| \leq \tilde{A}_{s,N}(0,0) = 1 .$$

For high signal-to-noise ratio, the "local accuracy" of the multipulse estimator can again be approximately characterized by the CRLBs. In the next theorem, proved in Appendix 8.B, the CRLBs are formulated under the simplifying assumption that the transmitted pulses $s_k(t)$ are either all real-valued or all conjugate even (see Appendix 8.B for a discussion of the general case).

Theorem 8.2 (CRLBs for Multipulse Range-Doppler Estimation) *Let the transmitted pulses $s_k(t)$ be all real-valued or all conjugate even ($s_k(-t) = s_k^*(t)$). Then, the variances of any unbiased joint estimator of the parameters τ_0 and ν_0 given the N received signals $r_k(t)$ in (8.9) are bounded as*

$$\mathrm{var}\{\hat{\tau}\} \geq \frac{-C_N}{\frac{d^2}{d\tau^2}|\tilde{A}_{s,N}(\tau,0)|^2\big|_{\tau=0}} , \qquad \mathrm{var}\{\hat{\nu}\} \geq \frac{-C_N}{\frac{d^2}{d\nu^2}|\tilde{A}_{s,N}(0,\nu)|^2\big|_{\nu=0}} ,$$

$$(8.12)$$

where $\tilde{A}_{s,N}(\tau,\nu)$ is the normalized AF of the transmitted pulse set $\{s_k(t)\}_{k=1}^N$ as defined in (8.11), and the SNR-dependent factor C_N is defined as

$$C_N = \frac{\eta}{\overline{E}_{r,N}}\left(1 + \frac{\eta}{\overline{E}_{r,N}}\right) , \qquad (8.13)$$

with the mean received energy

$$\overline{E}_{r,N} \triangleq 2\sigma_b^2 E_{s,N} .$$

These CRLBs can alternatively be expressed as

$$\mathrm{var}\{\hat{\tau}\} \geq \frac{C_N}{2(2\pi)^2}\frac{1}{\tilde{F}_{s,N}^2} , \qquad \mathrm{var}\{\hat{\nu}\} \geq \frac{C_N}{2(2\pi)^2}\frac{1}{\tilde{T}_{s,N}^2} , \qquad (8.14)$$

where $\tilde{T}_{s,N}$ and $\tilde{F}_{s,N}$, the average centered root-mean-square duration and bandwidth of the pulse set $\{s_k(t)\}_{k=1}^N$, are defined by (cf. (8.4)–(8.6))

$$\tilde{T}_{s,N}^2 \triangleq T_{s,N}^2 - t_{s,N}^2, \qquad \tilde{F}_{s,N}^2 \triangleq F_{s,N}^2 - f_{s,N}^2 \qquad (8.15)$$

with

$$T_{s,N}^2 = \frac{1}{E_{s,N}} \sum_{k=1}^N E_{s_k} T_{s_k}^2, \qquad t_{s,N} = \frac{1}{E_{s,N}} \sum_{k=1}^N E_{s_k} t_{s_k} \qquad (8.16)$$

and

$$F_{s,N}^2 = \frac{1}{E_{s,N}} \sum_{k=1}^N E_{s_k} F_{s_k}^2, \qquad f_{s,N} = \frac{1}{E_{s,N}} \sum_{k=1}^N E_{s_k} f_{s_k}. \qquad (8.17)$$

We note that the CRLB expressions in (8.14) follow from the CRLB expressions in (8.12) using the identities (see Appendix 8.B)

$$\left. \frac{d^2}{d\tau^2} \left|\tilde{A}_{s,N}(\tau,0)\right|^2 \right|_{\tau=0} = -2(2\pi)^2 \tilde{F}_{s,N}^2 \qquad (8.18)$$

$$\left. \frac{d^2}{d\nu^2} \left|\tilde{A}_{s,N}(0,\nu)\right|^2 \right|_{\nu=0} = -2(2\pi)^2 \tilde{T}_{s,N}^2. \qquad (8.19)$$

Similar to the single-pulse case (see Section 8.1.1), the CRLBs depend on the SNR $\overline{E}_{r,N}/\eta$ through C_N and on the transmitted pulse set $\{s_k(t)\}_{k=1}^N$ through the second partial derivatives (curvature) of $|\tilde{A}_{s,N}(\tau,\nu)|^2$ at the origin. Low CRLBs (i.e., potentially accurate estimates) require a sharp peak of $|\tilde{A}_{s,N}(\tau,\nu)|^2$ at the origin of the (τ,ν)-plane. Equivalently, (8.14) shows that good simultaneous range and Doppler estimation accuracy requires simultaneously large overall bandwidth and duration of the pulse set $\{s_k(t)\}_{k=1}^N$.

8.2.3 Global Accuracy

The analysis of the multipulse estimator's global accuracy is similarly analogous to the single-pulse case discussed in Section 8.1.2. The ML estimate being the location of the maximum of $\left|\sum_{k=1}^N A_{r_k,s_k}(\tau,\nu)\right|^2$, a simple measure of global accuracy is the expectation of this quantity. Again normalizing by the value at the true parameter point and in the absence of noise, we have

$$\kappa_{s,N}(\tau,\nu) \triangleq \frac{\mathrm{E}\left\{\left|\sum_{k=1}^N A_{r_k,s_k}(\tau,\nu)\right|^2\right\}}{\mathrm{E}\left\{\left|\sum_{k=1}^N A_{r_k,s_k}(\tau_0,\nu_0)\right|^2\right\}\Big|_{w_k(t)\equiv 0}}$$

$$= \left|\tilde{A}_{s,N}(\tau-\tau_0,\nu-\nu_0)\right|^2 + \frac{\eta}{\overline{E}_{r,N}}. \qquad (8.20)$$

For good global accuracy, $\kappa_{s,N}(\tau,\nu)$ must be small away from (τ_0,ν_0). This requires that the normalized auto-AS $|\tilde{A}_{s,N}(\tau,\nu)|^2$ of the transmitted pulse set $\{s_k(t)\}_{k=1}^N$ be small away from its peak at the origin of the (τ,ν)-plane.

Comparing the CRLBs in (8.12) or (8.14) and the function $\kappa_{s,N}(\tau,\nu)$ in (8.20) with the corresponding quantities (8.3), (8.7) and (8.8) obtained in Section 8.1 for the single-pulse case, we note that the results are strictly analogous with one major difference: the normalized auto-AF $\tilde{A}_{s,N}(\tau,\nu)$ of the pulse set $\{s_k(t)\}_{k=1}^N$ replaces the normalized auto-AF $\tilde{A}_s(\tau,\nu)$ of the transmitted pulse $s(t)$. In both the single-pulse case and the multipulse case, good global as well as local accuracy requires that the respective normalized auto-AS should have as much of a "thumbtack" shape as possible.

Furthermore, we note the obvious fact that for $N = 1$ the multipulse results reduce to the single-pulse results of Section 8.1.

8.2.4   Volume Bounds

Compared to the auto-AS of a single pulse, the auto-AS of N pulses has additional degrees of freedom that can result in a better thumbtack shape. With the peak height of $|\tilde{A}_{s,N}(\tau,\nu)|^2$ at the origin being $|\tilde{A}_{s,N}(0,0)|^2 = 1$ for all possible pulse sets $\{s_k(t)\}_{k=1}^N$, a simple global measure for the thumbtack shape is again the volume under $|\tilde{A}_{s,N}(\tau,\nu)|^2$,

$$V_{s,N} \triangleq \int_\tau \int_\nu \left| \tilde{A}_{s,N}(\tau,\nu) \right|^2 d\tau\, d\nu . \tag{8.21}$$

Our goal is the minimization of $V_{s,N}$. We recall that in the single-pulse case $V_{s,N}\big|_{N=1} = V_s = 1$ due to the radar uncertainty principle. The next theorem, shown in Appendix 8.C, characterizes the best case and the worst case.

Theorem 8.3 (AS Volume Bounds) *For given N, the AS volume $V_{s,N}$ is bounded as*

$$\frac{1}{N} \le V_{s,N} \le 1 . \tag{8.22}$$

The lower bound (corresponding to "globally optimum" pulses $s_k(t)$) is achieved if and only if the pulses $s_k(t)$ are orthogonal and have equal energies, i.e.,

$$V_{s,N} = V_{s,N}\big|_{\min} = \frac{1}{N} \qquad \Longleftrightarrow \qquad \langle s_k, s_l \rangle = E_s\,\delta_{kl} .$$

The upper bound (which corresponds to a worst-case choice of the $s_k(t)$ and equals the volume obtained in the single-pulse case $N = 1$) is achieved if and only if the pulses are all equal except for arbitrary complex factors, i.e.,

$$V_{s,N} = V_{s,N}\big|_{\max} = 1 \qquad \Longleftrightarrow \qquad s_k(t) = c_k\,s(t) ,$$

where $s(t)$ is an arbitrary signal.

Hence, the maximum AS volume reduction (as compared to the single-pulse case $N = 1$) is by a factor of N. This volume reduction can be interpreted as a relaxation of the radar uncertainty principle discussed in Section 8.1.3. In particular, we see that the AS volume can be made arbitrarily small if N, the number of transmitted pulses, is made sufficiently large.

8.3 Globally Optimum Pulse Sets and the Ambiguity Function of the Pulse Space

We shall now restrict our attention to orthogonal, equal-energy pulses, i.e., pulses satisfying $\langle s_k, s_l \rangle = E_s \, \delta_{kl}$.

8.3.1 The Case of Orthogonal, Equal-Energy Pulses

The case of orthogonal, equal-energy pulses can be considered as *globally optimum* since it corresponds to minimum AS volume[5]. Let

$$\mathcal{X} \triangleq \operatorname{span}\{s_k(t)\}_{k=1}^{N}$$

denote the space spanned by the orthogonal, equal-energy pulses $s_k(t)$. The dimension of the pulse space $\mathcal{X}$ equals N. The normalized pulses

$$x_k(t) \triangleq \frac{1}{\sqrt{E_s}} \, s_k(t)$$

form an orthonormal basis of the pulse space $\mathcal{X}$. It will be convenient to introduce the normalized AF of $\mathcal{X}$ as

$$\tilde{A}_{\mathcal{X}}(\tau, \nu) \triangleq \frac{1}{N} A_{\mathcal{X}}(\tau, \nu) = \frac{A_{\mathcal{X}}(\tau, \nu)}{A_{\mathcal{X}}(0, 0)}.$$

The normalized AF of the pulse set $\{s_k(t)\}_{k=1}^{N}$ is then equal to $\tilde{A}_{\mathcal{X}}(\tau, \nu)$:

$$\tilde{A}_{s,N}(\tau, \nu) = \frac{1}{E_{s,N}} \sum_{k=1}^{N} A_{s_k}(\tau, \nu) = \frac{E_s}{N E_s} \sum_{k=1}^{N} A_{x_k}(\tau, \nu) = \tilde{A}_{\mathcal{X}}(\tau, \nu),$$

where (7.3) has been used. This shows that the performance of the multipulse estimator (as measured by the shape of $|\tilde{A}_{s,N}(\tau, \nu)|^2$) is completely characterized by the AF of the pulse space $\mathcal{X}$; it does not otherwise depend on the specific pulses. Hence, any set of orthogonal, equal-energy pulses will achieve the same performance if the pulses span the same space and their energies are

---

[5]However, a *complete* characterization of estimator performance is not given by the volume under $|\tilde{A}_{s,N}(\tau, \nu)|^2$ but by the detailed shape of $|\tilde{A}_{s,N}(\tau, \nu)|^2$. In particular, the (minimum) AS volume should be distributed homogeneously over the effective AS support (outside the AS peak at the origin) in order to obtain small AS heights away from the origin.

equal. We note in passing that the minimum volume $V_{s,N}\big|_{\min} = \frac{1}{N}$ achieved by globally optimum pulse sets is consistent with the property (7.6) of the AF of a linear signal space. Furthermore, with (8.18), (8.19) the average centered root-mean-square duration and bandwidth that determine the CRLBs according to (8.14) can be written in terms of $\tilde{A}_\chi(\tau,\nu)$ as

$$\tilde{T}_{s,N}^2 = -\frac{1}{2(2\pi)^2}\frac{d^2}{d\nu^2}\big|\tilde{A}_\chi(0,\nu)\big|^2\bigg|_{\nu=0}$$

$$\tilde{F}_{s,N}^2 = -\frac{1}{2(2\pi)^2}\frac{d^2}{d\tau^2}\big|\tilde{A}_\chi(\tau,0)\big|^2\bigg|_{\tau=0} \ ,$$

and (8.20) becomes

$$\kappa_{s,N}(\tau,\nu) = \big|\tilde{A}_\chi(\tau-\tau_0,\nu-\nu_0)\big|^2 + \frac{\eta}{\overline{E}_{r,N}} \ .$$

Comparing (8.16), (8.17) with (3.9), it is also easily shown that for orthogonal, equal-energy pulses

$$T_{s,N} = T_\chi = -\frac{1}{(2\pi)^2}\frac{d^2}{d\nu^2}\tilde{A}_\chi(0,\nu)\bigg|_{\nu=0} \tag{8.23}$$

$$F_{s,N} = F_\chi = -\frac{1}{(2\pi)^2}\frac{d^2}{d\tau^2}\tilde{A}_\chi(\tau,0)\bigg|_{\tau=0} \ . \tag{8.24}$$

8.3.2 Hermite Spaces and Prolate Spheroidal Spaces Reconsidered

For the purposes of multipulse range-Doppler estimation, the "total space" $\mathcal{L}_2(\mathbb{R})$ of all finite-energy signals appears to be optimum since its AF has an infinitely sharp peak and zero sidelobe volume (cf. Section 7.3):

$$A_{\mathcal{L}_2(\mathbb{R})}(\tau,\nu) = \delta(\tau)\,\delta(\nu) \ .$$

Unfortunately, this space is not realistic for two reasons: (i) the dimension N is infinite, so that an infinite number of pulses has to be transmitted and processed; (ii) the pulses are distributed over the entire TF plane, i.e., the signal space is not limited with respect to time or frequency (recall that the WD of $\mathcal{L}_2(\mathbb{R})$ covers the entire TF plane).

In practice, we need finite-dimensional pulse spaces that are effectively time-limited and band-limited in the sense that all pulses decay rapidly outside a finite time interval and frequency band. Furthermore, the spaces should have good AF thumbtack shape, and it should be possible to adapt their dimension N as well as their effective time and frequency supports to practical specifications and limitations. The *Hermite spaces* previously considered in Sections 2.4,

3.3.2, 4.6.1, and 7.3 satisfy these requirements. We recall from Section 2.4 that the N-dimensional Hermite space $\mathcal{H}_N^{(T)}$ decays rapidly outside its effective time support $[-T_N, T_N]$ and frequency support $[-F_N, F_N]$, where $T_N = \sqrt{N/\pi}\, T$ and $F_N = \sqrt{N/\pi}\, \frac{1}{T}$. The time-scaling parameter T allows to trade off time duration against bandwidth or, equivalently, Doppler estimation accuracy against range estimation accuracy. In fact, the (centered) root-mean-square duration and bandwidth (see (8.15) and (8.23), (8.24)) of $\mathcal{H}_N^{(T)}$ can be shown to be

$$\tilde{T}^2_{\mathcal{H}_N^{(T)}} = T^2_{\mathcal{H}_N^{(T)}} = \frac{1}{4\pi} N T^2, \qquad \tilde{F}^2_{\mathcal{H}_N^{(T)}} = F^2_{\mathcal{H}_N^{(T)}} = \frac{1}{4\pi} \frac{N}{T^2}\,.$$

Thus, the CRLBs (see (8.14)) are obtained as

$$\mathrm{var}\{\hat{\tau}\} \geq \frac{C_N}{2\pi} \frac{T^2}{N}\,, \qquad \mathrm{var}\{\hat{\nu}\} \geq \frac{C_N}{2\pi} \frac{1}{NT^2}\,,$$

with C_N defined in (8.13). These results can be compared with those obtained for the individual Hermite functions $h_k^{(T)}(t)$. It can be shown that (cf. (8.4))

$$\tilde{T}^2_{h_k^{(T)}} = T^2_{h_k^{(T)}} = \frac{1}{4\pi} (2k-1)\, T^2, \qquad \tilde{F}^2_{h_k^{(T)}} = F^2_{h_k^{(T)}} = \frac{1}{4\pi} \frac{2k-1}{T^2}\,,$$

so that the CRLBs of a single-pulse estimator employing the kth Hermite function $h_k^{(T)}(t)$ are (see (8.7))

$$\mathrm{var}\{\hat{\tau}\} \geq \frac{C}{2\pi} \frac{T^2}{2k-1}\,, \qquad \mathrm{var}\{\hat{\nu}\} \geq \frac{C}{2\pi} \frac{1}{(2k-1)\, T^2}\,.$$

It is easily checked that $\tilde{T}^2_{\mathcal{H}_N^{(T)}}$ and $\tilde{F}^2_{\mathcal{H}_N^{(T)}}$ are the arithmetic averages of $\tilde{T}^2_{h_k^{(T)}}$ and $\tilde{F}^2_{h_k^{(T)}}$ $(k = 1, ..., N)$, respectively. Thus, the *peak* of the AS of $\mathcal{H}_N^{(T)}$ will be sharper than that of the AS of $h_1^{(T)}(t)$ but less sharp than that of the AS of $h_N^{(T)}(t)$. On the other hand, the *sidelobes* of the AS of $\mathcal{H}_N^{(T)}$ are much lower than those of the AS of any individual $h_k^{(T)}(t)$ (with the exception of $h_1^{(T)}(t)$ which does not have any sidelobes). A comparison between the AS of a Hermite space and that of individual Hermite functions has been given in Fig. 7.1 (see Section 7.3). We also recall that the perfect thumbtack shape achieved by the total space $\mathcal{L}_2(\mathbb{R})$ can be approximated arbitrarily closely if the dimension N of $\mathcal{H}_N^{(T)}$ (i.e., the number of Hermite pulses transmitted) is sufficiently large.

Logarithmic plots of AS cross sections for Hermite spaces and Hermite functions are shown in **Fig. 8.2**. (Note that, due to the elliptical symmetry of the AS, the AS shape is completely characterized by a cross section through

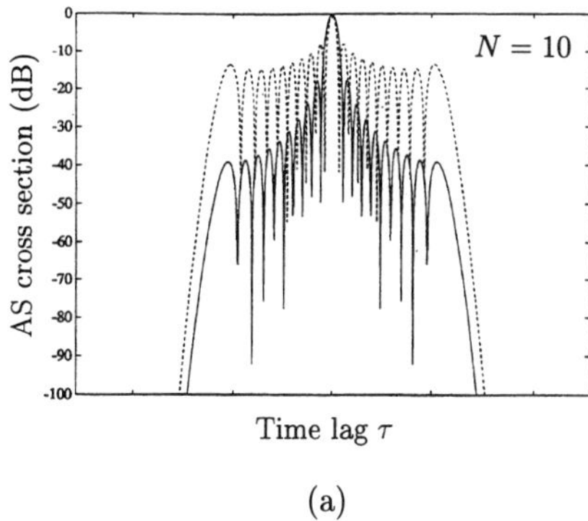

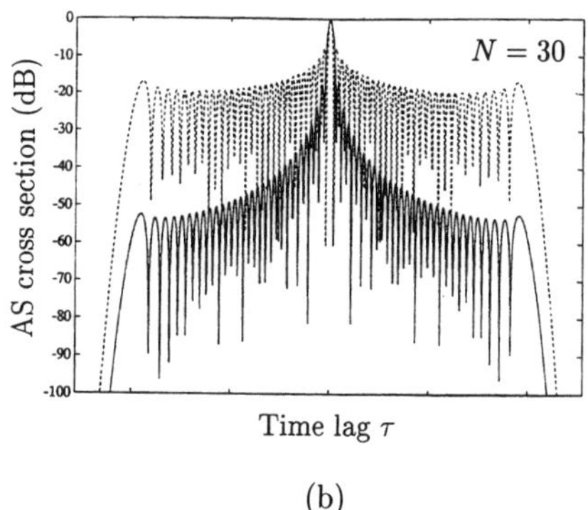

Figure 8.2. Cross sections of the normalized AS of the Nth Hermite function (broken line) and the normalized AS of the N-dimensional Hermite space (solid line) for (a) $N = 10$ and (b) $N = 30$.

the origin of the (τ, ν)-plane.) This comparison of "AS of a space versus AS of a signal" demonstrates the superiority of the multipulse estimator over the single-pulse estimator as far as AS sidelobe suppression is concerned. A further quantitative assessment of sidelobe reduction is given in **Fig. 8.3**.

The Hermite spaces are highly concentrated with respect to both time and frequency, but they are neither strictly time-limited nor strictly band-limited. In contrast, the prolate spheroidal spaces previously considered in Sections 2.4, 3.4.2, 4.6.2, and 7.3 are strictly bandlimited to a given frequency band $[-F/2, F/2]$ and optimally concentrated with respect to time. A comparison between the AS of a prolate spheroidal space and that of individual prolate spheroidal wave functions has been given in Fig. 7.2 (see Section 7.3).

Appendix 8.A: Proof of Theorem 8.1

In this appendix, we derive the multipulse ML range-Doppler estimator (8.10). We actually derive the multipulse ML estimator for a more general problem. Generalizing (8.9), we assume that our received signals are

$$r_k(t) = b\, s_k(t; \Theta) + w_k(t)\,, \qquad k = 1, ..., N\,, \qquad (8.A.1)$$

where Θ is a deterministic parameter or parameter vector to be estimated. For range-Doppler estimation, $\Theta = (\tau, \nu)$ and $s_k(t; \Theta) = s_k(t-\tau)\, e^{j2\pi\nu t}$. We assume that the energy of $s_k(t; \Theta)$ does not depend on Θ, $\int_t |s_k(t; \Theta)|^2\, dt = E_{s_k}$, which is satisfied in the range-Doppler case. As before, b is a zero-mean complex Gaussian random variable with $\mathrm{E}\{|b|^2\} = 2\sigma_b^2$. The $w_k(t)$ are zero-

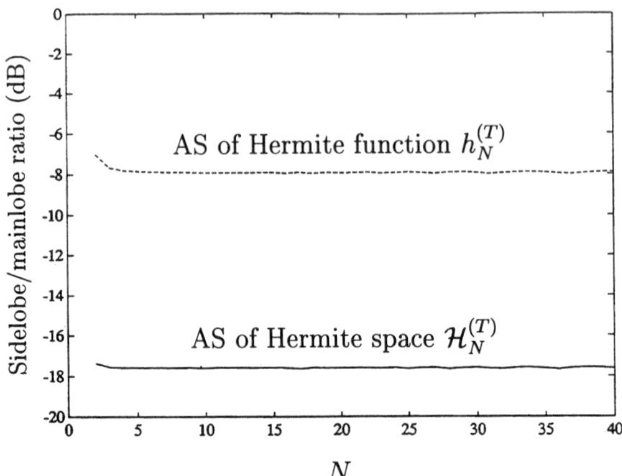

Figure 8.3. Ratio of maximum sidelobe height to main lobe (peak) height of the normalized AS of the Nth Hermite function (broken line) and the normalized AS of the N-dimensional Hermite space (solid line) for various dimensions N.

mean complex white Gaussian noise processes, each with power spectral density η, that are statistically independent of each other and also of b.

In order to derive the ML estimator of Θ, we first calculate the likelihood function, i.e., the probability density function (pdf) of the received data $r_k(t)$ indexed by Θ. To this purpose, we expand the $r_k(t)$ into an orthonormal basis $\{\phi_i(t)\}_{i=1}^M$ that is complete for $M \to \infty$. The expansion coefficients $r_{ki} = \langle r_k, \phi_i \rangle$ form a sufficient statistic for our estimation problem [Whalen, 1971]. It is convenient to define N vectors $\mathbf{r}_k$ of expansion coefficients r_{ki},

$$\mathbf{r}_k \overset{\triangle}{=} (r_{k1}, r_{k2}, ..., r_{kM})^T = b\,\mathbf{s}_k(\Theta) + \mathbf{w}_k\,, \qquad k = 1,...,N\,,$$

with $\mathbf{s}_k(\Theta) = \big(s_{k1}(\Theta), s_{k2}(\Theta), ..., s_{kM}(\Theta)\big)^T$ and $\mathbf{w}_k = (w_{k1}, w_{k2}, ..., w_{kM})^T$, $k = 1,...,N$, where the elements of the vectors $\mathbf{s}_k(\Theta)$ and $\mathbf{w}_k$ are $s_{ki}(\Theta) = \langle s_k(\,\cdot\,;\Theta), \phi_i \rangle$ and $w_{ki} = \langle w_k, \phi_i \rangle$. It is easily shown that, conditioned on b, the expansion coefficients r_{ki} are statistically independent Gaussian random variables with conditional mean $\mathrm{E}\{r_{ki}|b\} = b\,s_{ki}(\Theta)$ and conditional covariance $\mathrm{cov}\{r_{ki}, r_{lj}|b\} = \eta\,\delta_{kl}\,\delta_{ij}$. Hence, the conditional pdf of the vector $\mathbf{r}_k$ becomes

$$p_\Theta(\mathbf{r}_k|b) = \prod_{i=1}^M p_\Theta(r_{ki}|b) = \prod_{i=1}^M \frac{1}{\pi\eta}\,\exp\left(-\frac{1}{\eta}\left|r_{ki} - b\,s_{ki}(\Theta)\right|^2\right)$$

$$= \left(\frac{1}{\pi\eta}\right)^M \exp\left(-\frac{1}{\eta}\left\|\mathbf{r}_k - b\,\mathbf{s}_k(\Theta)\right\|^2\right)$$

where $\|\mathbf{x}\|^2 = \sum_i |x_i|^2$ and the subscript Θ in $p_\Theta(\mathbf{r}_k|b)$ indicates the pdf's dependence on the parameter Θ. We next combine all vectors $\mathbf{r}_k$ into a vector $\mathbf{r}$ of length NM,

$$\mathbf{r} \stackrel{\triangle}{=} (\mathbf{r}_1, \mathbf{r}_2, ..., \mathbf{r}_N)^T = b\,\mathbf{s}(\Theta) + \mathbf{w}$$

with $\mathbf{s}(\Theta) = \big(\mathbf{s}_1(\Theta), \mathbf{s}_2(\Theta), ..., \mathbf{s}_N(\Theta)\big)^T$ and $\mathbf{w} = (\mathbf{w}_1, \mathbf{w}_2, ..., \mathbf{w}_N)^T$. Conditioned on b, all component vectors $\mathbf{r}_k$ are statistically independent, Gaussian vectors with conditional mean $\mathrm{E}\{\mathbf{r}_k|b\} = b\,\mathbf{s}_k(\Theta)$ and conditional covariance $\mathrm{cov}\{\mathbf{r}_k,\mathbf{r}_l|b\} = \eta\,\mathbf{I}_M\,\delta_{kl}$, where $\mathbf{I}_M$ is the identity matrix of size M. It follows that the overall vector $\mathbf{r}$ has conditional mean $\mathrm{E}\{\mathbf{r}|b\} = b\,\mathbf{s}(\Theta)$ and conditional covariance $\mathrm{cov}\{\mathbf{r},\mathbf{r}\,|b\} = \eta\,\mathbf{I}_{NM}$. Hence, the conditional pdf of $\mathbf{r}$ becomes

$$p_\Theta(\mathbf{r}|b) = \prod_{k=1}^N p_\Theta(\mathbf{r}_k|b) = \prod_{k=1}^N \left(\frac{1}{\pi\eta}\right)^M \exp\left(-\frac{1}{\eta}\|\mathbf{r}_k - b\,\mathbf{s}_k(\Theta)\|^2\right)$$

$$= \left(\frac{1}{\pi\eta}\right)^{NM} \exp\left(-\frac{1}{\eta}\|\mathbf{r} - b\,\mathbf{s}(\Theta)\|^2\right).$$

With $\|\mathbf{r} - b\,\mathbf{s}(\Theta)\|^2 = \|\mathbf{r}\|^2 + |b|^2\,\|\mathbf{s}(\Theta)\|^2 - 2\,\mathrm{Re}\{b^*\,\mathbf{s}^H(\Theta)\,\mathbf{r}\}$, this can be rewritten as

$$p_\Theta(\mathbf{r}|b) = C(\mathbf{r})\,\exp\left(-\frac{1}{\eta}|b|^2\,\|\mathbf{s}(\Theta)\|^2\right)\exp\left(\frac{2}{\eta}\,\mathrm{Re}\{b^*\,\mathbf{s}^H(\Theta)\,\mathbf{r}\}\right),$$

where

$$C(\mathbf{r}) \stackrel{\triangle}{=} \left(\frac{1}{\pi\eta}\right)^{NM} \exp\left(-\frac{1}{\eta}\|\mathbf{r}\|^2\right)$$

is a factor that does not depend on Θ or b.

From $p_\Theta(\mathbf{r}|b)$, we now derive the unconditional pdf $p_\Theta(\mathbf{r})$ by integrating out the dependence on b. Converting to polar coordinates by letting $b = B\,e^{j\beta}$, the conditional pdf $p_\Theta(\mathbf{r}|b)$ becomes

$$p_\Theta(\mathbf{r}|B,\beta) = C(\mathbf{r})\,\exp\left(-\frac{1}{\eta}B^2\,\|\mathbf{s}(\Theta)\|^2\right)\exp\left(\frac{2B}{\eta}\,\mathrm{Re}\{e^{-j\beta}\,\mathbf{s}^H(\Theta)\,\mathbf{r}\}\right)$$

$$= C(\mathbf{r})\,\exp\left(-\frac{\|\mathbf{s}(\Theta)\|^2}{\eta}B^2\right)\exp\left(\frac{2\,G_\Theta(\mathbf{r})}{\eta}B\cos\big(\gamma_\Theta(\mathbf{r}) - \beta\big)\right)$$

where $G_\Theta(\mathbf{r})$ and $\gamma_\Theta(\mathbf{r})$ are the magnitude and phase, respectively, of $\mathbf{s}^H(\Theta)\,\mathbf{r}$, i.e.,

$$\mathbf{s}^H(\Theta)\,\mathbf{r} = G_\Theta(\mathbf{r})\,e^{j\gamma_\Theta(\mathbf{r})}. \tag{8.A.2}$$

The pdf of $b = B\,e^{j\beta}$ is $p(b) = p_1(B)\,p_2(\beta)$, where $p_1(B) = \frac{B}{\sigma_b^2}\exp\left(-\frac{B^2}{2\,\sigma_b^2}\right)$ $(B > 0)$ is a Rayleigh pdf with variance σ_b^2 and $p_2(\beta)$ is uniform over $[0, 2\pi)$. Integrating out the dependence on β yields [Whalen, 1971]

$$
\begin{aligned}
p_\Theta(\mathbf{r}|B) &= \int_\beta p_\Theta(\mathbf{r}|B,\beta)\,p_2(\beta)\,d\beta \\
&= C(\mathbf{r})\exp\left(-\frac{\|\mathbf{s}(\Theta)\|^2}{\eta}\,B^2\right)\frac{1}{2\pi}\int_0^{2\pi}\exp\left(\frac{2G_\Theta(\mathbf{r})}{\eta}\,B\,\cos\big(\gamma_\Theta(\mathbf{r})-\beta\big)\right)d\beta \\
&= C(\mathbf{r})\exp\left(-\frac{\|\mathbf{s}(\Theta)\|^2}{\eta}\,B^2\right)I_0\left(\frac{2\,G_\Theta(\mathbf{r})}{\eta}\,B\right)
\end{aligned}
$$

where $I_0(x) = \frac{1}{2\pi}\int_0^{2\pi} e^{x\cos\beta}d\beta$ is the modified zero-order Bessel function [Abramowitz and Stegun, 1965]. Integrating out the dependence on B next, the unconditional pdf is obtained as

$$
\begin{aligned}
p_\Theta(\mathbf{r}) &= \int_B p_\Theta(\mathbf{r}|B)\,p_1(B)\,dB \\
&= C(\mathbf{r})\int_0^\infty \exp\left(-\frac{\|\mathbf{s}(\Theta)\|^2}{\eta}\,B^2\right)I_0\left(\frac{2\,G_\Theta(\mathbf{r})}{\eta}\,B\right)\frac{B}{\sigma_b^2}\exp\left(-\frac{B^2}{2\sigma_b^2}\right)dB \\
&= C(\mathbf{r})\int_0^\infty \frac{B}{\sigma_b^2}\exp\left(-\frac{\eta + 2\sigma_b^2\,\|\mathbf{s}(\Theta)\|^2}{\eta}\,\frac{B^2}{2\sigma_b^2}\right)I_0\left(\frac{2\,G_\Theta(\mathbf{r})}{\eta}\,B\right)dB\,.
\end{aligned}
$$

According to [Whalen, 1971 (pp. 206-207)], this integral gives

$$
p_\Theta(\mathbf{r}) = C(\mathbf{r})\,K(\Theta)\,\exp\big[D(\Theta)\,G_\Theta^2(\mathbf{r})\big] \tag{8.A.3}
$$

with

$$
K(\Theta) = \frac{\eta}{\eta + 2\sigma_b^2\,\|\mathbf{s}(\Theta)\|^2}\,,\qquad
D(\Theta) = \frac{2\sigma_b^2}{\eta\,(\eta + 2\sigma_b^2\,\|\mathbf{s}(\Theta)\|^2)}\,. \tag{8.A.4}
$$

The ML estimator is obtained by maximizing $\ln p_\Theta(\mathbf{r})$ with respect to Θ [Whalen, 1971, Kay, 1993, Sorenson, 1980],

$$
\begin{aligned}
\hat\Theta_{\mathrm{ML}}(\mathbf{r}) &= \arg\max_\Theta\big\{\ln p_\Theta(\mathbf{r})\big\} = \arg\max_\Theta\big\{\ln C(\mathbf{r}) + \ln K(\Theta) + D(\Theta)\,G_\Theta^2(\mathbf{r})\big\} \\
&= \arg\max_\Theta\big\{\ln K(\Theta) + D(\Theta)\,G_\Theta^2(\mathbf{r})\big\}\,.
\end{aligned}
$$

For $M \to \infty$, we have

$$
\|\mathbf{s}(\Theta)\|^2 = \sum_{k=1}^N\left[\sum_{i=1}^\infty |s_{ki}|^2\right] = \sum_{k=1}^N E_{s_k} = E_{s,N}
$$

so that $K(\Theta) = K$ and $D(\Theta) = D \geq 0$ are independent of Θ (see (8.A.4)), and thus

$$\hat{\Theta}_{\mathrm{ML}}(\mathbf{r}) = \arg\max_{\Theta} \left\{ \ln K + D\, G_\Theta^2(\mathbf{r}) \right\} = \arg\max_{\Theta} G_\Theta^2(\mathbf{r})$$

with (cf. (8.A.2))

$$G_\Theta^2(\mathbf{r}) = \left| \mathbf{s}^H(\Theta)\,\mathbf{r} \right|^2 = \left| \sum_{k=1}^{N} \left[\sum_{i=1}^{\infty} s_{ki}^*(\Theta)\, r_{ki} \right] \right|^2 = \left| \sum_{k=1}^{N} \left[\int_t s_k^*(t;\Theta)\, r_k(t)\, dt \right] \right|^2 .$$

$$(8.A.5)$$

Hence, the final result for the ML estimator is

$$\hat{\Theta}_{\mathrm{ML}}(\mathbf{r}) = \arg\max_{\Theta} \left| \sum_{k=1}^{N} \chi_{r_k,s_k}(\Theta) \right|^2 \quad \text{with} \quad \chi_{r_k,s_k}(\Theta) = \int_t r_k(t)\, s_k^*(t;\Theta)\, dt .$$

$$(8.A.6)$$

In the special case of range-Doppler estimation, the kth signal is $s_k(t;\Theta) = s_k(t-\tau)\, e^{j2\pi\nu t}$ so that

$$\chi_{r_k,s_k}(\Theta) = \chi_{r_k,s_k}(\tau,\nu) = \int_t r_k(t)\, s_k^*(t-\tau)\, e^{-j2\pi\nu t}\, dt .$$

It is easily shown that $\chi_{r_k,s_k}(\tau,\nu) = e^{-j\pi\tau\nu}\, A_{r_k,s_k}(\tau,\nu)$, i.e., $\chi_{r_k,s_k}(\tau,\nu)$ equals the cross-AF $A_{r_k,s_k}(\tau,\nu)$ up to a phase factor. Thus, $\left| \sum_{k=1}^{N} \chi_{r_k,s_k}(\tau,\nu) \right|^2 = \left| \sum_{k=1}^{N} A_{r_k,s_k}(\tau,\nu) \right|^2$ and the ML estimator (8.A.6) becomes

$$(\hat{\tau},\hat{\nu})_{\mathrm{ML}} = \arg\max_{\tau,\nu} \left| \sum_{k=1}^{N} A_{r_k,s_k}(\tau,\nu) \right|^2 ,$$

which is the result (8.10) to be proved.

Appendix 8.B: Proof of Theorem 8.2

In order to derive the Cramér-Rao lower bounds (CRLBs) for the general multipulse setting (8.A.1), we first calculate the Fisher information matrix $\mathbf{J}$ whose elements are [Whalen, 1971, Kay, 1993, Sorenson, 1980]

$$J_{ij} = -\mathrm{E}\left\{ \frac{\partial^2}{\partial\theta_i\, \partial\theta_j} \ln p_\Theta(\mathbf{r}) \right\}\Bigg|_{\Theta=\Theta_0} .$$

Here, the θ_i are the scalar parameters contained in the parameter vector Θ and Θ_0 is the true value of Θ, i.e., the parameter vector actually contained in the received signals $r_k(t)$. Expressing $\ln p_\Theta(\mathbf{r})$ as in Appendix 8.A (see (8.A.3)-(8.A.6)), it can be shown that

$$J_{ij} = -D\,\mathrm{E}\left\{ \frac{\partial^2}{\partial\theta_i\,\partial\theta_j} \left| \sum_{k=1}^{N} \chi_{r_k,s_k}(\Theta) \right|^2 \right\}\Bigg|_{\Theta=\Theta_0} \qquad \text{with}\ \ D = \frac{2\sigma_b^2}{\eta\,(\eta + \overline{E}_{r,N})}.$$

With $\chi_{r_k,s_k}(\Theta) = \int_t r_k(t)\, s_k^*(t;\Theta)\, dt$ (see (8.A.6)), this becomes

$$J_{ij} = -D \sum_{k=1}^{N} \sum_{l=1}^{N} \int_{t_1} \int_{t_2} \mathrm{E}\{ r_k(t_1)\, r_l^*(t_2) \}$$

$$\cdot \left[\frac{\partial^2}{\partial\theta_i\,\partial\theta_j}\, s_k^*(t_1;\Theta)\, s_l(t_2;\Theta) \right]_{\Theta=\Theta_0} dt_1\, dt_2 . \qquad (8.\mathrm{B}.1)$$

Using $r_k(t) = b\, s_k(t;\Theta_0) + w_k(t)$ and the fact that the $w_k(t)$ are statistically independent of each other and of b, it is easily shown that $\mathrm{E}\{ r_k(t_1)\, r_l^*(t_2) \} = 2\sigma_b^2\, s_k(t_1;\Theta_0)\, s_l^*(t_2;\Theta_0) + \eta\, \delta_{kl}\, \delta(t_1 - t_2)$. Inserting this into (8.B.1) and doing some straightforward manipulations yields

$$J_{ij} = -D\, \frac{\partial^2}{\partial\theta_i\,\partial\theta_j} \left\{ 2\sigma_b^2 \left| \sum_{k=1}^{N} \chi_{s_k}(\Theta_0,\Theta) \right|^2 + \eta \sum_{k=1}^{N} \int_t |s_k(t;\Theta)|^2\, dt \right\}\Bigg|_{\Theta=\Theta_0}$$

where

$$\chi_{s_k}(\Theta_0,\Theta) \triangleq \int_t s_k(t;\Theta_0)\, s_k^*(t;\Theta)\, dt .$$

Since $\sum_{k=1}^{N} \int_t |s_k(t;\Theta)|^2\, dt = \sum_{k=1}^{N} E_{s_k} = E_{s,N}$ is independent of Θ, we obtain

$$J_{ij} = -D\, 2\sigma_b^2\, \frac{\partial^2}{\partial\theta_i\,\partial\theta_j} \left| \sum_{k=1}^{N} \chi_{s_k}(\Theta_0,\Theta) \right|^2 \Bigg|_{\Theta=\Theta_0} .$$

In the special case of range-Doppler estimation, we have $\Theta = (\tau,\nu)$, $\Theta_0 = (\tau_0,\nu_0)$, and $s_k(t;\Theta) = s_k(t-\tau)\, e^{j2\pi\nu t}$, and it easily follows that

$$\left| \sum_{k=1}^{N} \chi_{s_k}(\Theta_0,\Theta) \right|^2 = \left| \sum_{k=1}^{N} A_{s_k}(\tau-\tau_0,\nu-\nu_0) \right|^2 = E_{s,N}^2 \left| \tilde{A}_{s,N}(\tau-\tau_0,\nu-\nu_0) \right|^2 .$$

Hence, the Fisher information matrix $\mathbf{J}$ is a 2×2 matrix with elements

$$J_{ij} = -D\, 2\sigma_b^2\, E_{s,N}^2\, \frac{\partial^2}{\partial\theta_i\,\partial\theta_j} \left| \tilde{A}_{s,N}(\tau-\tau_0,\nu-\nu_0) \right|^2 \Bigg|_{\tau=\tau_0,\nu=\nu_0}$$

$$= -\frac{1}{C_N}\, \frac{\partial^2}{\partial\theta_i\,\partial\theta_j} \left| \tilde{A}_{s,N}(\tau,\nu) \right|^2 \Bigg|_{\tau=\nu=0} , \qquad 1 \le i,j \le 2$$

where

$$C_N = \frac{\eta}{\overline{E}_{r,N}} \left(1 + \frac{\eta}{\overline{E}_{r,N}} \right)$$

and $\theta_1 = \tau$ and $\theta_2 = \nu$. The CRLBs are derived from the inverse of $\mathbf{J}$ as [Whalen, 1971, Kay, 1993, Sorenson, 1980]

$$\mathrm{var}\{\hat{\tau}\} \geq (\mathbf{J}^{-1})_{11} = \frac{J_{22}}{J_{11}J_{22} - J_{12}J_{21}} = \frac{1}{1 - \gamma_{s,N}^2} \frac{-C_N}{\frac{\partial^2}{\partial\tau^2}|\tilde{A}_{s,N}(\tau,\nu)|^2\big|_{\tau=\nu=0}}$$

$$\mathrm{var}\{\hat{\nu}\} \geq (\mathbf{J}^{-1})_{22} = \frac{J_{11}}{J_{11}J_{22} - J_{12}J_{21}} = \frac{1}{1 - \gamma_{s,N}^2} \frac{-C_N}{\frac{\partial^2}{\partial\nu^2}|\tilde{A}_{s,N}(\tau,\nu)|^2\big|_{\tau=\nu=0}}$$

where

$$\gamma_{s,N} \triangleq \frac{\frac{\partial^2}{\partial\tau\partial\nu}|\tilde{A}_{s,N}(\tau,\nu)|^2\big|_{\tau=\nu=0}}{\sqrt{\frac{\partial^2}{\partial\tau^2}|\tilde{A}_{s,N}(\tau,\nu)|^2\big|_{\tau=\nu=0}\,\frac{\partial^2}{\partial\nu^2}|\tilde{A}_{s,N}(\tau,\nu)|^2\big|_{\tau=\nu=0}}}. \tag{8.B.2}$$

It will be shown presently that if the signals $s_k(t)$ are either all real-valued or all conjugate even, then $\frac{\partial^2}{\partial\tau\partial\nu}|\tilde{A}_{s,N}(\tau,\nu)|^2\big|_{\tau=\nu=0} = 0$ and thus $\gamma_{s,N} = 0$, so that the CRLBs reduce to

$$\mathrm{var}\{\hat{\tau}\} \geq \frac{-C_N}{\frac{\partial^2}{\partial\tau^2}|\tilde{A}_{s,N}(\tau,\nu)|^2\big|_{\tau=\nu=0}}, \qquad \mathrm{var}\{\hat{\nu}\} \geq \frac{-C_N}{\frac{\partial^2}{\partial\nu^2}|\tilde{A}_{s,N}(\tau,\nu)|^2\big|_{\tau=\nu=0}},$$
$$\tag{8.B.3}$$

which is the result (8.12) to be proved.

We next show the relations (8.18), (8.19) expressing the second AS derivatives (and in turn the CRLBs) in terms of $\tilde{T}_{s,N}$ and $\tilde{F}_{s,N}$. The derivation is simplified by introducing the normalized AF version

$$\tilde{\chi}_{s,N}(\tau,\nu) \triangleq \frac{1}{E_{s,N}} \sum_{k=1}^{N} \chi_{s_k}(\tau,\nu),$$

which equals the normalized AF $\tilde{A}_{s,N}(\tau,\nu)$ except for the fact that the AF version $\chi_{s_k}(\tau,\nu) = \int_t s_k(t)\, s_k^*(t-\tau)\, e^{-j2\pi\nu t}\, dt = e^{-j\pi\tau\nu} A_{s_k}(\tau,\nu)$ is used instead of $A_{s_k}(\tau,\nu)$. Since the two AFs are equal up to a phase factor, we have

$$\left|\tilde{A}_{s,N}(\tau,\nu)\right|^2 = \left|\tilde{\chi}_{s,N}(\tau,\nu)\right|^2. \tag{8.B.4}$$

We shall write $\mathcal{D}_{\theta_i}\{\cdot\} \triangleq \frac{\partial}{\partial\theta_i}\{\cdot\}\big|_{\tau=\nu=0}$ and $\mathcal{D}_{\theta_i\theta_j}\{\cdot\} \triangleq \frac{\partial^2}{\partial\theta_i\,\partial\theta_j}\{\cdot\}\big|_{\tau=\nu=0}$, so that the second AS derivatives at the origin can be written as $\mathcal{D}_{\theta_i\theta_j}\big|\tilde{A}_{s,N}\big|^2$.

Using (8.B.4) and carrying out the differentiation, we obtain

$$\mathcal{D}_{\theta_i\theta_j}|\tilde{A}_{s,N}|^2 \;=\; \mathcal{D}_{\theta_i\theta_j}\left\{\tilde{\chi}_{s,N}\,\tilde{\chi}^*_{s,N}\right\}$$

$$=\; 2\,\mathrm{Re}\left\{\mathcal{D}_{\theta_i\theta_j}\,\tilde{\chi}_{s,N} + \mathcal{D}_{\theta_i}\,\tilde{\chi}_{s,N}\left[\mathcal{D}_{\theta_j}\,\tilde{\chi}_{s,N}\right]^*\right\}. \qquad (8.B.5)$$

Since

$$\mathcal{D}_{\theta_i}\,\tilde{\chi}_{s,N} \;=\; \frac{1}{E_{s,N}}\sum_{k=1}^{N}\mathcal{D}_{\theta_i}\,\chi_{s_k}\,, \qquad \mathcal{D}_{\theta_i\theta_j}\,\tilde{\chi}_{s,N} \;=\; \frac{1}{E_{s,N}}\sum_{k=1}^{N}\mathcal{D}_{\theta_i\theta_j}\,\chi_{s_k}\,, \quad (8.B.6)$$

it remains to calculate $\mathcal{D}_{\theta_i}\,\chi_{s_k}$ and $\mathcal{D}_{\theta_i\theta_j}\,\chi_{s_k}$. With

$$\chi_{s_k}(\tau,\nu) \;=\; \int_t s_k(t)\,s_k^*(t-\tau)\,e^{-j2\pi\nu t}\,dt \;=\; \int_f S_k(f)\,S_k^*(f-\nu)\,e^{j2\pi\tau(f-\nu)}\,df\,,$$

we obtain after simple calculations

$$\mathcal{D}_\tau\,\chi_{s_k} \;=\; \int_f S_k(f)\,S_k^*(f-\nu)\,\frac{\partial}{\partial\tau}\left\{e^{j2\pi\tau(f-\nu)}\right\}df\,\bigg|_{\tau=\nu=0} \;=\; j2\pi E_{s_k}f_{s_k}$$

$$\mathcal{D}_\nu\,\chi_{s_k} \;=\; \int_t s_k(t)\,s_k^*(t-\tau)\,\frac{\partial}{\partial\nu}\left\{e^{-j2\pi\nu t}\right\}dt\,\bigg|_{\tau=\nu=0} \;=\; -j2\pi E_{s_k}t_{s_k}$$

$$\mathcal{D}_{\tau\tau}\,\chi_{s_k} \;=\; \int_f S_k(f)\,S_k^*(f-\nu)\,\frac{\partial^2}{\partial\tau^2}\left\{e^{j2\pi\tau(f-\nu)}\right\}df\,\bigg|_{\tau=\nu=0} \;=\; -(2\pi)^2 E_{s_k}F^2_{s_k}$$

$$\mathcal{D}_{\nu\nu}\,\chi_{s_k} \;=\; \int_t s_k(t)\,s_k^*(t-\tau)\,\frac{\partial^2}{\partial\nu^2}\left\{e^{-j2\pi\nu t}\right\}dt\,\bigg|_{\tau=\nu=0} \;=\; -(2\pi)^2 E_{s_k}T^2_{s_k}$$

$$\mathcal{D}_{\tau\nu}\,\chi_{s_k} \;=\; \int_t s_k(t)\,\frac{\partial^2}{\partial\tau\partial\nu}\left\{s_k^*(t-\tau)\,e^{-j2\pi\nu t}\right\}dt\,\bigg|_{\tau=\nu=0}$$

$$=\; \int_t s_k(t)\int_f S_k^*(f)\,\frac{\partial}{\partial\tau}\left\{e^{-j2\pi(t-\tau)f}\right\}\frac{\partial}{\partial\nu}\left\{e^{-j2\pi\nu t}\right\}dt\,df\,\bigg|_{\tau=\nu=0}$$

$$=\; (2\pi)^2 E_{s_k}X_{s_k}\,,$$

where t_{s_k}, f_{s_k}, $T^2_{s_k}$, and $F^2_{s_k}$ have been defined in (8.5) and (8.6), and

$$X_{s_k} \;\triangleq\; \frac{1}{E_{s_k}}\int_t\int_f tf\,s_k(t)\,S_k^*(f)\,e^{-j2\pi tf}\,dt\,df\,.$$

Inserting into (8.B.6) yields

$$\mathcal{D}_\tau \, \tilde{\chi}_{s,N} \;=\; j2\pi f_{s,N} \tag{8.B.7}$$

$$\mathcal{D}_\nu \, \tilde{\chi}_{s,N} \;=\; -j2\pi t_{s,N} \tag{8.B.8}$$

$$\mathcal{D}_{\tau\tau} \, \tilde{\chi}_{s,N} \;=\; -(2\pi)^2 F_{s,N}^2 \tag{8.B.9}$$

$$\mathcal{D}_{\nu\nu} \, \tilde{\chi}_{s,N} \;=\; -(2\pi)^2 \, T_{s,N}^2 \tag{8.B.10}$$

$$\mathcal{D}_{\tau\nu} \, \tilde{\chi}_{s,N} \;=\; (2\pi)^2 X_{s,N} \, , \tag{8.B.11}$$

where $t_{s,N}$, $f_{s,N}$, $T_{s,N}^2$, and $F_{s,N}^2$ have been defined in (8.16) and (8.17), and

$$X_{s,N} \;\overset{\triangle}{=}\; \frac{1}{E_{s,N}} \sum_{k=1}^{N} E_{s_k} X_{s_k} \, .$$

Inserting (8.B.7)–(8.B.11) into (8.B.5) gives

$$
\begin{aligned}
\mathcal{D}_{\tau\tau} \left| \tilde{A}_{s,N} \right|^2 \;&=\; 2\,\mathrm{Re}\left\{ \mathcal{D}_{\tau\tau}\, \tilde{\chi}_{s,N} + \left| \mathcal{D}_\tau \, \tilde{\chi}_{s,N} \right|^2 \right\} \\[4pt]
&=\; -2(2\pi)^2 \left(F_{s,N}^2 - f_{s,N}^2 \right) \;=\; -2(2\pi)^2 \tilde{F}_{s,N}^2 \\[8pt]
\mathcal{D}_{\nu\nu} \left| \tilde{A}_{s,N} \right|^2 \;&=\; 2\,\mathrm{Re}\left\{ \mathcal{D}_{\nu\nu}\, \tilde{\chi}_{s,N} + \left| \mathcal{D}_\nu \, \tilde{\chi}_{s,N} \right|^2 \right\} \\[4pt]
&=\; -2(2\pi)^2 \left(T_{s,N}^2 - t_{s,N}^2 \right) \;=\; -2(2\pi)^2 \tilde{T}_{s,N}^2 \\[8pt]
\mathcal{D}_{\tau\nu} \left| \tilde{A}_{s,N} \right|^2 \;&=\; 2\,\mathrm{Re}\left\{ \mathcal{D}_{\tau\nu}\, \tilde{\chi}_{s,N} + \mathcal{D}_\tau \, \tilde{\chi}_{s,N} \left[\mathcal{D}_\nu \, \tilde{\chi}_{s,N} \right]^* \right\} \\[4pt]
&=\; 2(2\pi)^2 \left(\mathrm{Re}\{X_{s,N}\} - t_{s,N} f_{s,N} \right)
\end{aligned}
$$

which completes the proof of (8.18), (8.19).

We finally show that $\mathcal{D}_{\tau\nu}|\tilde{A}_{s,N}|^2 = 0$ if the pulses $s_k(t)$ are either all real-valued or all conjugate even. For real-valued $s_k(t)$, there is $S_k(-f) = S_k^*(f)$ from which it follows (i) that $f_{s_k} = 0$ and hence $f_{s,N} = 0$, and (ii) that $\mathrm{Re}\{ \int_f f\, S_k^*(f)\, e^{-j2\pi tf}\, df \} = 0$. This gives

$$\mathrm{Re}\{X_{s_k}\} \;=\; \frac{1}{E_{s_k}} \int_t t\, s_k(t)\, \mathrm{Re}\left\{ \int_f f\, S_k^*(f)\, e^{-j2\pi tf}\, df \right\} dt \;=\; 0$$

and further $\mathrm{Re}\{X_{s,N}\} = 0$. Combined with $f_{s,N} = 0$, we have $\mathcal{D}_{\tau\nu}|\tilde{A}_{s,N}|^2 = 2(2\pi)^2 \left(\mathrm{Re}\{X_{s,N}\} - t_{s,N} f_{s,N} \right) = 0$. An analogous proof shows that $\mathcal{D}_{\tau\nu}|\tilde{A}_{s,N}|^2 = 0$ if all $s_k(t)$ are conjugate even, i.e., if $s_k(-t) = s_k^*(t)$. Inserting $\mathcal{D}_{\tau\nu}|\tilde{A}_{s,N}|^2 = 0$ into (8.B.2) yields $\gamma_{s,N} = 0$, and thus gives the simplified CRLBs in (8.B.3).

Appendix 8.C: Proof of Theorem 8.3

We want to show the bounds (8.22) on the volume $V_{s,N}$. Inserting $\tilde{A}_{s,N}(\tau,\nu) = \frac{1}{E_{s,N}} \sum_{k=1}^{N} A_{s_k}(\tau,\nu)$ into (8.21), interchanging summations and integrations, and using Moyal's formula (1.11), the AS volume becomes

$$V_{s,N} = \frac{1}{E_{s,N}^2} \sum_{k=1}^{N} \sum_{l=1}^{N} \int_\tau \int_\nu A_{s_k}(\tau,\nu)\, A_{s_l}^*(\tau,\nu)\, d\tau\, d\nu = \frac{1}{E_{s,N}^2} \sum_{k=1}^{N} \sum_{l=1}^{N} |\langle s_k, s_l \rangle|^2.$$

$$(8.\text{C}.1)$$

Lower Bound. The double sum in (8.C.1) can be split up and bounded from below as

$$V_{s,N} = \frac{1}{E_{s,N}^2} \left[\sum_{k=1}^{N} \|s_k\|^4 + \sum_{\substack{k=1 \\ k \neq l}}^{N} \sum_{l=1}^{N} |\langle s_k, s_l \rangle|^2 \right]$$

$$\geq \frac{1}{E_{s,N}^2} \sum_{k=1}^{N} \|s_k\|^4 = \frac{1}{E_{s,N}^2} \sum_{k=1}^{N} E_{s_k}^2 , \qquad (8.\text{C}.2)$$

with equality if and only if $\langle s_k, s_l \rangle = 0$ for $k \neq l$, i.e., if all signals $s_k(t)$ are orthogonal. Next, we minimize this lower bound with respect to the energies E_{s_k} under the side constraint of fixed total transmitted energy, i.e., $E_{s,N} = \sum_{k=1}^{N} E_{s_k} = E_0$ where E_0 is a constant. Using a Lagrange multiplier λ, this amounts to the unconstrained minimization of

$$v(\{E_{s_k}\}) \triangleq \frac{1}{E_0^2} \sum_{k=1}^{N} E_{s_k}^2 + \lambda \left(\sum_{k=1}^{N} E_{s_k} - E_0 \right).$$

Setting the derivatives of $v(\{E_{s_k}\})$ with respect to the E_{s_k} equal to zero yields the system of equations $E_{s_k} = -E_0^2 \lambda/2$, which shows that all E_{s_k} must be equal, i.e., $E_{s_k} = E_{s,N}/N$. Inserting this into (8.C.2), the (absolutely) minimum volume is then obtained as

$$V_{s,N}\Big|_{\text{min}} = \frac{1}{E_{s,N}^2} \sum_{k=1}^{N} \left(\frac{E_{s,N}}{N} \right)^2 = \frac{1}{N} ,$$

which is the lower bound to be proved. Note that our derivation has also shown that this lower bound is attained if and only if all $s_k(t)$ are orthogonal with equal energies.

Upper Bound. Using Schwarz' inequality $|\langle s_k, s_l \rangle|^2 \leq \|s_k\|^2 \|s_l\|^2 = E_{s_k} E_{s_l}$ in (8.C.1), we obtain

$$V_{s,N} \leq \frac{1}{E_{s,N}^2} \sum_{k=1}^{N} \sum_{l=1}^{N} E_{s_k} E_{s_l} = \frac{1}{E_{s,N}^2} E_{s,N} E_{s,N} = 1$$

with equality if and only if $s_k(t) = c_{kl}\, s_l(t)$ or, equivalently, $s_k(t) = c_k\, s(t)$ with an arbitrary signal $s(t)$.

9 CONCLUSIONS

We have proposed and developed a time-frequency (TF) analysis of linear signal spaces which is primarily based on two novel "TF space representations" called the *Wigner distribution of a linear signal space* and the *ambiguity function of a linear signal space*. These TF space representations are extensions of the Wigner distribution (WD) of a signal and the ambiguity function (AF) of a signal, respectively, to linear signal spaces.

In a certain sense, these extensions are deceptively trivial: the WD of a space is simply the sum of the WDs of all orthonormal basis signals spanning the space, and similarly for the AF. However, the WD of a space is also the Weyl symbol of the space's orthogonal projection operator (which is independent of a particular basis), and it has been shown to satisfy a number of interesting properties that extend well-known properties satisfied by the WD of a signal. Again, similar arguments apply to the AF of a space.

What is gained by a TF analysis of signal spaces? First of all, a signal space ceases to be merely an abstract (if useful) mathematical concept. Using the WD or AF, a linear signal space is displayed, or visualized, as a surface over a joint TF plane, with the shape of this surface indicating important properties of the space or, equivalently, of the associated projection operator. From the

WD, we can derive important space parameters such as the space's dimension (energy), effective duration, or effective bandwidth in a simple and intuitively appealing manner. Moreover, some fundamental linear transformations of a space become very simple geometrical TF coordinate transformations when viewed through the WD. The AF, on the other hand, visualizes the *TF shifts* that may occur when a signal is projected onto the space. Hence, the WD and AF are useful analysis tools capable of showing important properties of signal spaces, and facilitating an intuitive understanding of signal spaces.

Apart from pure analysis and visualization applications, there exist other applications that are related to specific signal processing tasks. The WD-based *synthesis* of signal spaces has been shown in Chapter 5 to allow the systematic and, in a specific sense, optimum design of *TF projection filters* that pass a given TF region and suppress the rest of the TF plane. Such filters are well suited for the separation of signal components located in known, effectively disjoint TF regions. A related application of WD-based space synthesis is the construction of *TF basis systems* that allow the parsimonious expansion of signals located in known TF regions. The application of TF projection filters and TF signal expansions in a stochastic signal estimation and signal detection context has been considered in Chapter 6. The AF, on the other hand, has been shown in Chapter 8 to be closely related to a maximum-likelihood multipulse procedure for estimating the range and Doppler shift of a slowly fluctuating point target—a parameter estimation problem that is important in radar and sonar.

The concepts and results presented in this work can be extended in several directions. These extensions also provide some suggestions for future research.

- While the WD and the AF have been emphasized, we have also formulated a general method for redefining any other quadratic TF signal representation for linear signal spaces. Various quadratic TF signal representations are known to be specifically suited to certain types of signals—for example, the *Altes-Marinovich Q-distribution* [Altes, 1990, Marinovich, 1986, Papandreou et al., 1993] is specifically suited to hyperbolically frequency-modulated signals. Using our general method, it is straightforward to redefine these quadratic TF signal representations for spaces. What remains to be done is discuss the properties of these new TF space representations and characterize the spaces for which they have specific advantages.

- An important limitation of the TF projection filters, TF expansions, and TF filter banks considered in Chapter 5 is that they are only useful for an "off-line" mode of signal processing. Indeed, these methods are based on an eigendecomposition of a matrix whose size equals (or is proportional to) the signal length. This entails a severe practical limitation of the signal length and also is incompatible with an on-line mode of operation. A "short-time version" of the TF projection filter that circumvents this limitation can easily be formulated. However, here the resulting system no longer corresponds to a true orthogonal projection operator. Other on-line, short-time methods for TF filtering are based on the short-time Fourier transform, Gabor expansion, wavelet transform, or other signal expansions effecting a tiling of the TF plane. However, the TF resolution of this latter type of on-line filters is not as good as that of a TF projection filter, and the TF pass regions that can be realized are more or less restricted, depending on the specific signal expansion used. Hence, finding computationally efficient on-line methods for TF filtering that preserve the complete generality of TF pass regions and have excellent TF resolution (two major advantages of the TF projection filter) is still an open problem.

- The WD of a space can alternatively be viewed as a TF representation of any orthonormal basis spanning the space. An extension to *non-orthogonal* function sets, specifically *frames* [Daubechies, 1990, Duffin and Schaeffer, 1952, Heil and Walnut, 1989], has been proposed recently [Hlawatsch and Bölcskei, 1994]. Frames are of fundamental relevance to important signal expansions (such as the Gabor expansion and the wavelet transform) and to the theory of irregular sampling [Benedetto, 1992]. Generalizing the WD of a linear signal space to frames leads to *TF frame representations* that show how a frame's numerical properties depend on the TF location of the signal to be expanded. TF frame representations thus yield a much more detailed characterization of a frame than the frame bounds. In some cases (e.g., Weyl-Heisenberg frames), they also facilitate an adjustment of certain frame parameters with the aim of improving a frame's numerical properties [Hlawatsch and Bölcskei, 1994]. Since frames and signal expansions are often mathematically equivalent to *filter bank analysis/synthesis systems* [Cvetković, 1995, Bölcskei et al., 1995, Bölcskei et al., 1996, Bölcskei and Hlawatsch, 1998, Janssen, 1998], these TF frame representations can also be applied to filter banks.

- Furthermore, the WD of a space can alternatively be viewed as a TF representation of the space's orthogonal projection operator. The extension of this concept to more general linear, time-varying (LTV) systems leads to two distinct TF representations of LTV systems called the *Weyl symbol of an LTV system* [Kozek and Hlawatsch, 1991b, Kozek, 1992b, Kozek and Hlawatsch, 1992, Kozek, 1992a, Shenoy and Parks, 1994, Folland, 1989, Janssen, 1989] and the *WD of an LTV system* [Hlawatsch, 1992b, Hlawatsch and Kozek, 1992]. For "underspread" LTV systems, i.e., systems that do not introduce substantial TF shifts, these TF representations show the "TF weighting characteristic" of the LTV system. A special case of an underspread system is the orthogonal projection operator on a *simple* (i.e., non-sophisticated) signal space (see Subsections 3.1.2 and 7.4.1).

Thus, we conclude this work by looking beyond the TF analysis and synthesis of linear signal spaces and pointing at a more general principle: that of generalizing TF *signal* analysis and synthesis to other mathematical quantities important in signal processing, such as frames, filter banks, and LTV systems. We believe that this principle may lead to interesting new developments.

References

Abramowitz, M. and Stegun, I. (1965). *Handbook of Mathematical Functions.* Dover, New York.

Altes, R. A. (1980). Detection, estimation and classification with spectrograms. *J. Acoust. Soc. Amer.*, 67(4):1232–1246.

Altes, R. A. (1990). Wide-band, proportional-bandwidth Wigner-Ville analysis. *IEEE Trans. Acoust., Speech, Signal Processing*, 38(6):1005–1012.

Baker, C. R. (1966). Optimum quadratic detection of a random vector in Gaussian noise. *IEEE Trans. Comm. Technol.*, 14:802–805.

Bello, P. A. (1963). Characterization of randomly time-variant linear channels. *IEEE Trans. Comm. Syst.*, 11:360–393.

Benedetto, J. J. (1992). Irregular sampling and frames. In Chui, C. K., editor, *Wavelets: A Tutorial in Theory and Applications*, pages 445–507. Academic Press, New York.

Bertrand, J. and Bertrand, P. (1992a). Affine time-frequency distributions. In Boashash, B., editor, *Time-Frequency Signal Analysis – Methods and Applications*, pages 118–140. Longman Cheshire, Melbourne.

Bertrand, J. and Bertrand, P. (1992b). A class of affine Wigner functions with extended covariance properties. *J. Math. Phys.*, 33(7):2515–2527.

Boashash, B. (1990). Time-frequency signal analysis. In Haykin, S., editor, *Advances in Spectrum Estimation*, volume 1, pages 418–517. Prentice Hall, Englewood Cliffs (NJ).

Bölcskei, H. and Hlawatsch, F. (1998). Oversampled modulated filter banks. In Feichtinger, H. G. and Strohmer, T., editors, *Gabor Analysis and Algorithms: Theory and Applications*, pages 295–322. Birkhäuser, Boston (MA).

Bölcskei, H., Hlawatsch, F., and Feichtinger, H. G. (1995). Equivalence of DFT filter banks and Gabor expansions. In *Proc. SPIE Wavelet Applications in*

Signal and Image Processing III, volume 2569, Part I, pages 128–139, San Diego (CA).

Bölcskei, H., Hlawatsch, F., and Feichtinger, H. G. (1996). Frame-theoretic analysis and design of oversampled filter banks. In *Proc. IEEE ISCAS-96*, volume 2, pages 409–412, Atlanta (GA).

Boudreaux-Bartels, G. F. (1997). Time-varying signal processing using Wigner distribution synthesis techniques. In Mecklenbräuker, W. and Hlawatsch, F., editors, *The Wigner Distribution — Theory and Applications in Signal Processing*, pages 269–317. Elsevier, Amsterdam (The Netherlands).

Boudreaux-Bartels, G. F. and Parks, T. W. (1986). Time-varying filtering and signal estimation using Wigner distribution synthesis techniques. *IEEE Trans. Signal Processing*, 34:442–451.

Bourdier, R., Allard, J. F., and Trumpf, K. (1988). Effective frequency response and signal replica generation for filtering algorithms using multiplicative modifications of the STFT. *Signal Processing*, 15(2):193–201.

Cimini, L. J. and Kassam, S. A. (1981). Optimum piecewise-constant Wiener filters. *J. Opt. Soc.*, 71(10):1162–1171.

Claasen, T. A. C. M. and Mecklenbräuker, W. F. G. (1980). The Wigner distribution — A tool for time-frequency signal analysis; Part I: Continuous-time signals. *Philips J. Research*, 35(3):217–250.

Claasen, T. A. C. M. and Mecklenbräuker, W. F. G. (1980a). The Wigner distribution — A tool for time-frequency signal analysis; Part II: Discrete-time signals. *Philips J. Research*, 35(4/5):276–300.

Claasen, T. A. C. M. and Mecklenbräuker, W. F. G. (1980b). The Wigner distribution — A tool for time-frequency signal analysis; Part III: Relations with other time-frequency signal transformations. *Philips J. Research*, 35(6):372–389.

Claasen, T. A. C. M. and Mecklenbräuker, W. F. G. (1984). On the time-frequency discriminiation of energy distributions: Can they look sharper than Heisenberg? In *Proc. IEEE ICASSP-84*, pages 41B7.1–41B7.4, San Diego (CA).

Cohen, L. (1995). *Time-Frequency Analysis*. Prentice Hall, Englewood Cliffs (NJ).

Cook, C. E. and Bernfeld, M. (1993). *Radar Signals—An Introduction to Theory and Applications*. Artech House, Norwood (MA).

Cover, T. M. and Thomas, J. A. (1991). *Elements of Information Theory*. Wiley, New York.

Cvetković, Z. (1995). Oversampled modulated filter banks and tight Gabor frames in $l^2(\mathbb{Z})$. In *Proc. IEEE ICASSP-95*, pages 1456–1459, Detroit (MI).

Daubechies, I. (1988). Time-frequency localization operators: A geometric phase space approach. *IEEE Trans. Inf. Theory*, 34(4):605–612.

Daubechies, I. (1990). The wavelet transform, time-frequency localization and signal analysis. *IEEE Trans. Inf. Theory*, 36:961–1005.

Daubechies, I. and Paul, T. (1988). Time-frequency localization operators: A geometric phase space approach—II. The use of dilations. *Inverse Problems*, 4:661–680.

de Bruijn, N. G. (1967). Uncertainty principles in Fourier analysis. In Shisha, O., editor, *Inequalities*, pages 57–71. Academic Press, New York.

Duffin, R. J. and Schaeffer, A. C. (1952). A class of nonharmonic Fourier series. *Trans. Amer. Math. Soc.*, 72:341–366.

Duijvelaar, W. J. (1984). Asymptotic behaviour of some Laguerre series. Technical Report 5986, Philips Research Laboratories, Eindhoven, The Netherlands.

Farkash, S. and Raz, S. (1994). Linear systems in Gabor time-frequency space. *IEEE Trans. Signal Processing*, 42(3):611–617.

Flandrin, P. (1988). Maximum signal energy concentration in a time-frequency domain. In *Proc. IEEE ICASSP-88*, pages 2176–2179, New York.

Flandrin, P. (1989). Time-dependent spectra for nonstationary stochastic processes. In Longo, G. and Picinbono, B., editors, *Time and Frequency Representation of Signal and Systems*, pages 69–124. Springer, Wien.

Flandrin, P. (1993). *Temps-fréquence*. Hermès, Paris.

Flandrin, P. and Gonçalvès, P. (1994). From wavelets to time-scale energy distributions. In Schumaker, L. L. and Webb, G., editors, *Recent Advances in Wavelet Analysis*, pages 309–334. Academic Press.

Flandrin, P. and Gonçalvès, P. (1996). Geometry of affine time-frequency distributions. *Applied and Computational Harmonic Analysis*, 3:10–39.

Flandrin, P. and Martin, W. (1997). The Wigner-Ville spectrum of nonstationary random signals. In Mecklenbräuker, W. and Hlawatsch, F., editors, *The Wigner Distribution — Theory and Applications in Signal Processing*, pages 211–267. Elsevier, Amsterdam (The Netherlands).

Folland, G. B. (1989). *Harmonic Analysis in Phase Space*, volume 122 of *Annals of Mathematics Studies*. Princeton University Press, Princeton (NJ).

Folland, G. B. and Sitaram, A. (1997). The uncertainty principle: A mathematical survey. *J. Fourier Anal. Appl.*, 3(3):207–238.

Franks, L. E. (1969). *Signal Theory*. Prentice Hall, Englewood Cliffs (NJ).

Haykin, S. (1991). *Adaptive Filter Theory*. Prentice Hall, Englewood Cliffs (NJ).

Heil, C., Ramanathan, J., and Topiwala, P. (1994). Phase space localization operators: Asymptotics of singular values. In *Proc. IEEE-SP Int. Sympos. Time-Frequency Time-Scale Analysis*, pages 194–196, Philadelphia (PA).

Heil, C. E. and Walnut, D. F. (1989). Continuous and discrete wavelet transforms. *SIAM Rev.*, 31(4):628–666.

Hlawatsch, F. (1991). Duality and classification of bilinear time-frequency signal representations. *IEEE Trans. Signal Processing*, 39(7):1564–1574.

Hlawatsch, F. (1992a). Regularity and unitarity of bilinear time-frequency signal representations. *IEEE Trans. Inf. Theory*, 38(1):82–94.

Hlawatsch, F. (1992b). Wigner distribution analysis of linear, time-varying systems. In *Proc. IEEE ISCAS-92*, pages 1459–1462, San Diego (CA).

Hlawatsch, F. and Bölcskei, H. (1994). Time-frequency analysis of frames. In *Proc. IEEE-SP Int. Sympos. Time-Frequency Time-Scale Analysis*, pages 52–55, Philadelphia (PA).

Hlawatsch, F. and Boudreaux-Bartels, G. F. (1992). Linear and quadratic time-frequency signal representations. *IEEE Signal Processing Magazine*, 9(2):21–67.

Hlawatsch, F., Costa, A. H., and Krattenthaler, W. (1994). Time-frequency signal synthesis with time-frequency extrapolation and don't-care regions. *IEEE Trans. Signal Processing*, 42(9):2513–2520.

Hlawatsch, F. and Edelson, G. S. (1992). The ambiguity function of a linear signal space and its application to maximum likelihood range/Doppler estimation. In *Proc. IEEE-SP Int. Sympos. Time-Frequency Time-Scale Analysis*, pages 489–492, Victoria, Canada.

Hlawatsch, F. and Flandrin, P. (1997). The interference structure of the Wigner distribution and related time-frequency signal representations. In Mecklenbräuker, W. and Hlawatsch, F., editors, *The Wigner Distribution — Theory and Applications in Signal Processing*, pages 59–133. Elsevier, Amsterdam (The Netherlands).

Hlawatsch, F. and Kozek, W. (1991). Time-frequency analysis of linear signal spaces. In *Proc. IEEE ICASSP-91*, pages 2045–2048, Toronta, Canada.

Hlawatsch, F. and Kozek, W. (1992). Time-frequency weighting and displacement effects in linear, time-varying systems. In *Proc. IEEE ISCAS-92*, pages 1455–1458, San Diego (CA).

Hlawatsch, F. and Kozek, W. (1993). The Wigner distribution of a linear signal space. *IEEE Trans. Signal Processing*, 41(3):1248–1258.

Hlawatsch, F. and Kozek, W. (1994). Time-frequency projection filters and time-frequency signal expansions. *IEEE Trans. Signal Processing*, 42(12):3321–3334.

Hlawatsch, F. and Kozek, W. (1995). Second-order time-frequency synthesis of nonstationary random processes. *IEEE Trans. Inf. Theory*, 41(1):255–267.

Hlawatsch, F. and Krattenthaler, W. (1989). A new approach to time-frequency signal decomposition. In *Proc. IEEE ISCAS-89*, pages 1248–1251, Portland (OR).

Hlawatsch, F. and Krattenthaler, W. (1992). Bilinear signal synthesis. *IEEE Trans. Signal Processing*, 40(2):352–363.

Hlawatsch, F. and Krattenthaler, W. (1997). Signal synthesis algorithms for bilinear time-frequency signal representations. In Mecklenbräuker, W. and Hlawatsch, F., editors, *The Wigner Distribution — Theory and Applications in Signal Processing*, pages 135–209. Elsevier, Amsterdam (The Netherlands).

Hlawatsch, F., Krattenthaler, W., and Kozek, W. (1990). Time-frequency subspaces and their application to time-varying filtering. In *Proc. IEEE ICASSP-90*, pages 1607–1610, Albuquerque (NM).

Hlawatsch, F., Papandreou, A., and Boudreaux-Bartels, G. F. (1993a). The power classes of quadratic time-frequency representations: A generalization of the affine and hyperbolic classes. In *Proc. 27th Asilomar Conf. Signals, Systems, Computers*, pages 1265–1270, Pacific Grove (CA).

Hlawatsch, F., Papandreou, A., and Boudreaux-Bartels, G. F. (1993b). Regularity and unitarity of affine and hyperbolic time-frequency representations. In *Proc. IEEE ICASSP-93*, pages 245–248, Minneapolis (MN).

Hlawatsch, F., Papandreou, A., and Boudreaux-Bartels, G. F. (1997). The hyperbolic class of quadratic time-frequency representations—Part II: Subclasses, intersection with the affine and power classes, regularity, and unitarity. *IEEE Trans. Signal Processing*, 45(2):303–315.

Hlawatsch, F. and Urbanke, R. (1994). Bilinear time-frequency representations of signals: The shift-scale invariant class. *IEEE Trans. Signal Processing*, 42(2):357–366.

Honig, M. L. and Messerschmitt, D. G. (1984). *Adaptive Filters: Structures, Algorithms, and Applications*. Kluwer, Boston (MA).

Janssen, A. J. E. M. (1982). On the locus and spread of pseudo-density functions in the time-frequency plane. *Philips J. Research*, 37(3):79–110.

Janssen, A. J. E. M. (1989). Wigner weight functions and Weyl symbols of non-negative definite linear operators. *Philips J. Research*, 44:7–42.

Janssen, A. J. E. M. (1997a). *Private communication*.

Janssen, A. J. E. M. (1997b). Positivity and spread of bilinear time-frequency distributions. In Mecklenbräuker, W. and Hlawatsch, F., editors, *The Wigner Distribution — Theory and Applications in Signal Processing*, pages 1–58. Elsevier, Amsterdam (The Netherlands).

Janssen, A. J. E. M. (1998). The duality condition for Weyl-Heisenberg frames. In Feichtinger, H. G. and Strohmer, T., editors, *Gabor Analysis and Algorithms: Theory and Applications*, pages 33–84. Birkhäuser, Boston (MA).

Kadambe, S. and Boudreaux-Bartels, G. F. (1992). A comparison of the existence of 'cross terms' in the Wigner distribution and the squared magnitude of the wavelet transform and the short-time Fourier transform. *IEEE Trans. Signal Processing*, 40(10):2498–2517.

Kay, S. M. (1993). *Fundamentals of Statistical Signal Processing: Estimation Theory*. Prentice Hall, Englewood Cliffs (NJ).

Klauder, J. R. (1960). The design of radar signals having both high range resolution and high velocity resolution. *Bell Syst. Tech. J.*, 39(4):809–820.

Kozek, W. (1992a). On the generalized Weyl correspondence and its application to time-frequency analysis of linear time-varying systems. In *Proc. IEEE-SP Int. Sympos. Time-Frequency Time-Scale Analysis*, pages 167–170, Victoria, B.C., Canada.

Kozek, W. (1992b). Time-frequency signal processing based on the Wigner-Weyl framework. *Signal Processing*, 29(1):77–92.

Kozek, W. (1996). On the underspread/overspread classification of nonstationary random processes. In Kirchgässner, K., Mahrenholtz, O., and Mennicken, R., editors, *Proc. Int. Conf. Industrial and Applied Mathematics (Hamburg, July 1995)*, volume 3 of *Mathematical Research*, pages 63–66, Berlin. Akademieverlag.

Kozek, W. (1997a). *Matched Weyl-Heisenberg Expansions of Nonstationary Environments*. PhD thesis, Vienna University of Technology.

Kozek, W. (1997b). On the transfer function calculus for underspread LTV channels. *IEEE Trans. Signal Processing*, 45(1):219–223.

Kozek, W. (1998). Adaptation of Weyl-Heisenberg frames to underspread environments. In Feichtinger, H. G. and Strohmer, T., editors, *Gabor Analysis and Algorithms: Theory and Applications*, pages 323–352. Birkhäuser, Boston (MA).

Kozek, W. and Hlawatsch, F. (1991a). Time-frequency filter banks with perfect reconstruction. In *Proc. IEEE ICASSP-91*, pages 2049–2052, Toronto, Canada.

Kozek, W. and Hlawatsch, F. (1991b). Time-frequency representation of linear time-varying systems using the Weyl symbol. In *Proc. 6th Int. Conf. on Digital Processing of Signals in Communication*, pages 25–30, Loughborough, UK.

Kozek, W. and Hlawatsch, F. (1992). A comparative study of linear and nonlinear time-frequency filters. In *Proc. IEEE-SP Int. Sympos. Time-Frequency Time-Scale Analysis*, pages 163–166, Victoria, B.C., Canada.

Kozek, W., Hlawatsch, F., Kirchauer, H., and Trautwein, U. (1994). Correlative time-frequency analysis and classification of nonstationary random processes. In *Proc. IEEE-SP Int. Sympos. Time-Frequency Time-Scale Analysis*, pages 417–420, Philadelphia (PA).

Krattenthaler, W. and Hlawatsch, F. (1993). Time-frequency design and processing of signals via smoothed Wigner distributions. *IEEE Trans. Signal Processing*, 41(1):278–287.

Lafrance, P. (1990). *Fundamental Concepts in Communication*. Prentice Hall, Englewood Cliffs (NJ).

Landau, H. J. and Pollak, H. O. (1961). Prolate spheroidal wave functions, Fourier analysis and uncertainty—II. *Bell System Technical Journal*, 40(1):65–84.

Lawson, C. L. and Hanson, R. J. (1974). *Solving Least Squares Problems*. Prentice Hall, Englewood Cliffs, NJ.

Lee, E. A. and Messerschmitt, D. G. (1994). *Digital Communication*. Kluwer, Boston (MA), 2nd edition.

Luenberger, D. G. (1969). *Optimization by Vector Space Methods*. Wiley, New York.

Marinovich, N. M. (1986). *The Wigner distribution and the ambiguity function: Generalizations, enhancement, compression and some applications*. PhD thesis, City University of New York.

Martin, W. and Flandrin, P. (1985). Wigner-Ville spectral analysis of nonstationary processes. *IEEE Trans. Acoust., Speech, Signal Processing*, 33(6):1461–1470.

Naylor, A. W. and Sell, G. R. (1982). *Linear Operator Theory in Engineering and Science*. Springer, New York, 2nd edition.

Papandreou, A., Hlawatsch, F., and Boudreaux-Bartels, G. F. (1993). The hyperbolic class of quadratic time-frequency representations—Part I: Constant-Q warping, the hyperbolic paradigm, properties, and members. *IEEE Trans. Signal Processing*, 41(12):3425–3444.

Papandreou, A., Hlawatsch, F., and Boudreaux-Bartels, G. F. (1995). A unified framework for the scale covariant affine, hyperbolic, and power class quadratic time-frequency representations using generalized time shifts. In *Proc. IEEE ICASSP-95*, pages 1017–1020, Detroit (MI).

Papoulis, A. (1984a). *Probability, Random Variables, and Stochastic Processes*. McGraw-Hill, New York.

Papoulis, A. (1984b). *Signal Analysis*. McGraw-Hill, Singapore.

Parks, T. W. and Shenoy, R. G. (1990). Time-frequency concentrated basis functions. In *Proc. IEEE ICASSP-90*, volume V, pages 2459–2462, Albuquerque (NM).

Picinbono, B. C. (1980). A geometrical interpretation of signal detection and estimation. *IEEE Trans. Inf. Theory*, 26(4):493–497.

Picinbono, B. C. and Duvaut, P. (1988). Optimal linear-quadratic systems for detection and estimation. *IEEE Trans. Inf. Theory*, 34(2):304–311.

Poor, H. V. (1988). *An Introduction to Signal Detection and Estimation*. Springer, New York.

Portnoff, M. R. (1980). Time-frequency representation of digital signals and systems based on short-time Fourier analysis. *IEEE Trans. Acoust., Speech, Signal Processing*, 28(1):55–69.

Price, R. and Hofstetter, E. M. (1965). Bounds on the volume and height distributions of the ambiguity function. *IEEE Trans. Inf. Theory*, 11:207–214.

Qian, S. and Chen, D. (1996). *Joint Time-Frequency Analysis*. Prentice Hall, Englewood Cliffs (NJ).

Ramanathan, J. and Topiwala, P. (1993). Time-frequency localization via the Weyl correspondence. *SIAM J. Matrix Anal. Appl.*, 24(5):1378–1393.

Rihaczek, A. W. (1969). *Principles of High-Resolution Radar*. McGraw Hill, New York.

Rioul, O. and Flandrin, P. (1992). Time-scale energy distributions: A general class extending wavelet transforms. *IEEE Trans. Signal Processing*, 40(1):1746–1757.

Saleh, B. E. A. and Subotic, N. S. (1985). Time-variant filtering of signals in the mixed time-frequency domain. *IEEE Trans. Acoust., Speech, Signal Processing*, 33(6):1479–1485.

Scharf, L. L. (1991). *Statistical Signal Processing*. Addison Wesley, Reading (MA).

Shenoy, R. G. and Parks, T. W. (1994). The Weyl correspondence and time-frequency analysis. *IEEE Trans. Signal Processing*, 42(2):318–331.

Skolnik, M. I. (1980). *Introduction to Radar Systems*. McGraw-Hill, New York.

Slepian, D. and Pollak, H. O. (1961). Prolate spheroidal wave functions, Fourier analysis and uncertainty—I. *Bell System Technical Journal*, 40(1):43–64.

Sorenson, H. W. (1980). *Parameter Estimation: Principles and Problems*. Marcel Dekker, New York.

Sostrand, K. A. (1968). Mathematics of the time-varying channel. *Proc. NATO Advanced Study Inst. on Signal Processing with Emphasis on Underwater Acoustics*, 2:25.1–25.20.

Szegö, G. (1975). *Orthogonal Polynomials*. AMS Colloquium Publications, New York, 4th edition.

Szu, H. H. and Blodgett, J. A. (1981). Wigner distribution and ambiguity function. In Narducci, L. M., editor, *Optics in Four Dimensions*, pages 355–381. American Inst. of Physics, New York.

Therrien, C. W. (1992). *Discrete Random Signals and Statistical Signal Processing*. Prentice Hall, Englewood Cliffs (NJ).

Umesh, S. and Tufts, D. W. (1992). Resolving the components of transient signals by a multistage procedure. In *Proc. IEEE ICASSP-92*, pages 553–556, San Francisco (CA).

Van Trees, H. L. (1968). *Detection, Estimation, and Modulation Theory, Part I: Detection, Estimation, and Linear Modulation Theory*. Wiley, New York.

Van Trees, H. L. (1992). *Detection, Estimation, and Modulation Theory: Radar-Sonar Signal Processing and Gaussian Signals in Noise*. Krieger, Malabar (FL).

Whalen, A. D. (1971). *Detection of Signals in Noise.* Academic Press, New York.

Wigner, E. P. (1932). On the quantum correction for thermodynamic equilibrium. *Phys. Rev.*, 40:749–759.

Wilcox, C. H. (1991). The synthesis problem for radar ambiguity functions. In Blahut, R. E., Miller, Jr., W., and Wilcox, C. H., editors, *Radar and Sonar*, pages 229–260. Springer, New York.

Woodward, P. M. (1953). *Probability and Information Theory with Application to Radar.* Pergamon Press, London.

Wozencraft, J. M. and Jacobs, I. M. (1965). *Principles of Communication Engineering.* Wiley, New York.

Zadeh, L. A. (1950). Frequency analysis of variable networks. *Proc. of IRE*, 76:291–299.

Index